Building Bluetooth Low Energy Systems

Take your first steps in IoT

Muhammad Usama bin Aftab

BIRMINGHAM - MUMBAI

Building Bluetooth Low Energy Systems

First published: April 2017

Production reference: 1200417

Published by Packt Publishing Ltd.
Livery Place
35 Livery Street
Birmingham
B3 2PB, UK.
ISBN 978-1-78646-108-7

www.packtpub.com

Credits

Author
Muhammad Usama bin Aftab

Reviewer
Gustavo Litovsky

Commissioning Editor
Kartikey Pandey

Acquisition Editors
Rahul Nair
Sonali Vernekar

Content Development Editor
Amedh Gemraram Pohad

Technical Editor
Vishal Kamal Mewada

Copy Editor
Madhusudan Uchil

Project Coordinator
Virginia Dias

Proofreader
Safis Editing

Indexer
Pratik Shirodkar

Graphics
Kirk D'Penha

Production Coordinator
Shantanu Zagade

About the Author

Muhammad Usama bin Aftab is a software engineer at one of the leading solar lighting manufacturers in North America. At Carmanah Technologies, he works closely with the product development and management teams to develop software for embryonic technologies such as Bluetooth Low Energy. He also focuses on full-stack development using Angular 2 and Java. His main areas of interest range from mobile application development to information security. Mr. Aftab is an alumnus of the University of Victoria, Canada, where he received his master's degree in applied science in electrical and computer engineering. Mr. Aftab has experience in Android, iOS, and cross-platform gaming technologies (such as Unity 3D), which led him to coauthor *Learning Android Intents* for Packt in 2014.

I am using this opportunity to express thanks to my parents, to whom I owe all my gratitude. Not only in academia but in every walk of life, they supported me. Thank you, mom and dad, for everything you did for me.

I would like to extend my thanks to Fanie Collardeau, who not only motivated me but guided me in tough times. She made me a better person and gave me a totally different yet wonderful perspective on life.

I am grateful to Ronald Schouten, who provided me an open and flexible environment to work in. His persona brought a positive attitude in my professional life and I sincerely thank him for being such an amazing mentor.

Thanks also to Darren Beckwith, who gave me the motivation to learn and adopt new technologies (including Bluetooth Low Energy). He understood the importance of my book and gave me encouragement to pursue writing.

Last but not least, thank you Carmanah Technologies, who laid the foundation of my knowledge in the Internet of Things and provided me an incredible environment to work in.

Among the hills, when you sit in the cool shade of the white poplars, sharing the peace and serenity of distant fields and meadows - then let your heart say in silence, "God rests in reason." And when the storm comes, and the mighty wind shakes the forest, and thunder and lightning proclaim the majesty of the sky - then let your heart say in awe, "God moves in passion." And since you are a breath in God's sphere and a leaf in God's forest, you too should rest in reason and move in passion.

- Kahlil Gibran

About the Reviewer

Gustavo Litovsky is a lifelong engineer and expert on Bluetooth and Wi-Fi. As the founder and CEO of Argenox Technologies, he leads a team of engineers in building amazing connected products for some of the world's most recognizable brands as well as numerous startups. Prior to founding Argenox, he spent time consulting at Samsung and several other companies on integrating wireless connectivity. He previously worked as an engineer at Texas Instrument's Wireless Connectivity business unit, supporting customers with Wi-Fi, Bluetooth, and GPS solutions for products such as the Nike FuelBand, Pebble Smartwatch, Motorola Droid Bionic, among many others. He is also an inventor and co-inventor of several patents.

I would like to thank my wife, Rebecca, whose help and support has been crucial in everything I do. I am grateful also to my parents and children, for all their love and support.

www.PacktPub.com

For support files and downloads related to your book, please visit www.PacktPub.com.

Did you know that Packt offers eBook versions of every book published, with PDF and ePub files available? You can upgrade to the eBook version at www.PacktPub.com and as a print book customer, you are entitled to a discount on the eBook copy. Get in touch with us at service@packtpub.com for more details.

At www.PacktPub.com, you can also read a collection of free technical articles, sign up for a range of free newsletters and receive exclusive discounts and offers on Packt books and eBooks.

https://www.packtpub.com/mapt

Get the most in-demand software skills with Mapt. Mapt gives you full access to all Packt books and video courses, as well as industry-leading tools to help you plan your personal development and advance your career.

Why subscribe?

- Fully searchable across every book published by Packt
- Copy and paste, print, and bookmark content
- On demand and accessible via a web browser

Customer Feedback

Thanks for purchasing this Packt book. At Packt, quality is at the heart of our editorial process. To help us improve, please leave us an honest review on this book's Amazon page at https://www.amazon.com/dp/1786461080.

If you'd like to join our team of regular reviewers, you can e-mail us at customerreviews@packtpub.com. We award our regular reviewers with free eBooks and videos in exchange for their valuable feedback. Help us be relentless in improving our products!

Table of Contents

Preface

Classic Bluetooth became a revolutionary technology when it replaced line-of-sight infrared communication. It was almost unbelievable to imagine that you could walk in close proximity and still transfer files from one cell phone to another. Classic Bluetooth broke the conventions and penetrated deep into the consumer electronics market.

When it comes to technology, weaknesses and mistakes are not easily tolerated. The Bluetooth SIG understands this, which is why they rethought the very foundation of classic Bluetooth. With the launch of Bluetooth Low Energy, they stamped their claim over not just small-distance communication, but also considerably long-distance communication using meshes. Providing the low-power, low-cost, and high-distance communication of Bluetooth 5 will surely benefit BLE in the Internet of Things.

Building Bluetooth Low Energy Systems is a book that understands the importance of this technology. It discusses the background of Bluetooth Low Energy and how it is being used in different IoT products. This book provides a practical understanding of the fundamentals of communication: the client/server architecture of Bluetooth. It does not stop there--it moves forward to explain Bluetooth beacons with indoor navigation, Bluetooth CSRMesh, and Bluetooth-Internet gateways. All these compelling implementations of BLE are not just explained but put into code using different APIs. With these examples, the book will bring a deeper understanding of Bluetooth Low Energy systems as they exist today.

Changing technology brings motivation to learners and gives them encouragement to apply it in their daily lives. This book will provide you with necessary tools to learn and apply this technology. It will give you enough knowledge about Bluetooth Low Energy that you will move many steps ahead in developing stuff. It is highly encouraged that you learn from the book and build advance Bluetooth Low Energy systems.

With the launch of Bluetooth 5, the SIG made known their intentions of making this technology a front-line player in the Internet of Things. We already have Bluetooth sitting in our pockets; it is just a matter of time that everything around is connected to us through it. This is why it is worthwhile to spend time learning this fast-growing and continuously expanding technology.

What this book covers

Chapter 1, *BLE and the Internet of Things*, is a review of the basics for those familiar with BLE. If you are new to BLE, then this chapter will get you up to speed for future chapters. This chapter will speak about the emerging domain of this decade with many real-life examples where we've been using Internet of Things (IoT) services without knowing it. It will also provide examples of wearables, autonomous vehicle, and sports gear to make you understand the potential of IoT in the coming years.

Chapter 2, *BLE Hardware, Software, and Debugging Tools*, gives an overview of the BLE devices available in the market. It will provide a comparison and detailed description of the different hardware with their corresponding SDKs. It will also explain the working of some applications available for Android/iOS and for desktop computers.

Chapter 3, *Building a BLE Central and Peripheral Communication System*, provides examples to understand communication in BLE using central and peripherals. Step-by-step explanations of the examples will help you understand the working of BLE. Java will be used to illustrate Android code.

Chapter 4, *Bluetooth Low Energy Beacons*, gives an introduction to Bluetooth beacon technology. Classification of BLE beacons and their role in the IoT will be explained with examples. This chapter will also contain the two types of beacons currently available, EddyStone and iBeacon. It will also provide example applications using Eddystone and iBeacons on Estimote Beacons and Tags.

Chapter 5, *BLE Indoor Navigation Using Estimote Beacons*, focuses on an important aspect of BLE beacons: using them as an indoor navigation system.

Chapter 6, *Bluetooth Mesh Technology*, teaches you about the mesh technology introduced by CSRMesh. This chapter will focus on the key aspects of mesh technology, such as flooding, routing, and security considerations. The chapter will also help you build a CSRMesh control application using Android.

Chapter 7, *Implementing a Bluetooth Gateway Using the Raspberry Pi 3, talks about the importance of gateway technology to connect Bluetooth-enabled devices to the internet. It will also demonstrate an example application to create a gateway that will act as a bridge between BLE-enabled devices and the Internet using the Raspberry Pi 3.*

Chapter 8, *The Future of Bluetooth Low Energy*, talks about the future of the Internet of Things. It will discuss the new version of BLE (version 5.0). It will also talk about the potential applications of IoT in the future, with smart cities and machine-to-machine communication. The future of IoT in industrial development will also be discussed in this chapter.

What you need for this book

In order to keep up with the book, you will need to spend some time learning basic coding. In addition to RESTful APIs, this book refers to three major programming languages:

- Java 8 for Android
- SWIFT 3 for iOS
- Node.js

Additionally, you will need following software tools:

- Android Studio 2.3
- XCode 8
- The nRF Connect Android/iOS app.
- Estimote apps and SDK for Android and iOS
- CSR uEnergy tools
- CSRMesh Library for Android
- NOOBS version 2.4.0
- Bluetooth Developer Studio

From a hardware standpoint, this book will require you to work with:

- CSRMesh Development Kit
- Raspberry Pi 3 with Bluetooth Low Energy.
- iPhone with BLE to implement iOS applications
- Android phone with BLE to implement Android applications
- Estimote Location Beacons

This book also refers to the Estimote Beacon Management dashboard in the cloud to manage your beacons.

Who this book is for

The book is for developers and enthusiasts who are passionate about learning Bluetooth Low Energy technologies and want to add new features and services to their new or existing products. They should be familiar with programming languages such as Swift, Java, and JavaScript. Knowledge of debugging skills would be an advantage.

Conventions

In this book, you will find a number of text styles that distinguish between different kinds of information. Here are some examples of these styles and an explanation of their meaning.

Code words in text, database table names, folder names, filenames, file extensions, pathnames, dummy URLs, user input, and Twitter handles are shown as follows: "The next lines of code read the link and assign it to the to the `BeautifulSoup` function."

A block of code is set as follows:

```
<uses-permission android:name="android.permission.BLUETOOTH" />
<uses-permission android:name="android.permission.BLUETOOTH_ADMIN" />
```

When we wish to draw your attention to a particular part of a code block, the relevant lines or items are set in bold:
[default]
<uses-permission android:name="**android.permission.BLUETOOTH**" />
<uses-permission android:name="**android.permission.BLUETOOTH_ADMIN**" />

New terms and **important words** are shown in bold. Words that you see on the screen, for example, in menus or dialog boxes, appear in the text like this: "Select the menu bar on the left and select **Configure GATT Server**."

Warnings or important notes appear in a box like this.

Tips and tricks appear like this.

Reader feedback

Feedback from our readers is always welcome. Let us know what you think about this book-what you liked or disliked. Reader feedback is important for us as it helps us develop titles that you will really get the most out of.

To send us general feedback, simply e-mail feedback@packtpub.com, and mention the book's title in the subject of your message.

If there is a topic that you have expertise in and you are interested in either writing or contributing to a book, see our author guide at www.packtpub.com/authors.

Customer support

Now that you are the proud owner of a Packt book, we have a number of things to help you to get the most from your purchase.

Downloading the example code

You can download the example code files for this book from your account at http://www.packtpub.com. If you purchased this book elsewhere, you can visit http://www.packtpub.com/support and register to have the files e-mailed directly to you.

You can download the code files by following these steps:

1. Log in or register to our website using your e-mail address and password.
2. Hover the mouse pointer on the **SUPPORT** tab at the top.
3. Click on **Code Downloads & Errata**.
4. Enter the name of the book in the **Search** box.
5. Select the book for which you're looking to download the code files.

6. Choose from the drop-down menu where you purchased this book from.
7. Click on Code Download.

Once the file is downloaded, please make sure that you unzip or extract the folder using the latest version of:

- WinRAR / 7-Zip for Windows
- Zipeg / iZip / UnRarX for Mac
- 7-Zip / PeaZip for Linux

The code bundle for the book is also hosted on GitHub at `https://github.com/PacktPubl ishing/Building-Bluetooth-Low-Energy-Systems`. We also have other code bundles from our rich catalog of books and videos available at `https://github.com/PacktPublish ing/`. Check them out!

Downloading the color images of this book

We also provide you with a PDF file that has color images of the screenshots/diagrams used in this book. The color images will help you better understand the changes in the output. You can download this file from `https://www.packtpub.com/sites/default/files/down loads/BuildingBluetoothLowEnergySystems_ColorImages.pdf`.

Errata

Although we have taken every care to ensure the accuracy of our content, mistakes do happen. If you find a mistake in one of our books-maybe a mistake in the text or the code-we would be grateful if you could report this to us. By doing so, you can save other readers from frustration and help us improve subsequent versions of this book. If you find any errata, please report them by visiting `http://www.packtpub.com/submit-errata`, selecting your book, clicking on the **Errata Submission Form** link, and entering the details of your errata. Once your errata are verified, your submission will be accepted and the errata will be uploaded to our website or added to any list of existing errata under the Errata section of that title.

To view the previously submitted errata, go to `https://www.packtpub.com/books/conten t/support` and enter the name of the book in the search field. The required information will appear under the **Errata** section.

Piracy

Piracy of copyrighted material on the Internet is an ongoing problem across all media. At Packt, we take the protection of our copyright and licenses very seriously. If you come across any illegal copies of our works in any form on the Internet, please provide us with the location address or website name immediately so that we can pursue a remedy.

Please contact us at copyright@packtpub.com with a link to the suspected pirated material.

We appreciate your help in protecting our authors and our ability to bring you valuable content.

Questions

If you have a problem with any aspect of this book, you can contact us at questions@packtpub.com, and we will do our best to address the problem.

1
BLE and the Internet of Things

This book is a practical guide to the world of the **Internet of Things** (**IoT**), where you will not only learn the theoretical concepts of the Internet of Things but also will get a number of practical examples. The purpose of this book is to bridge the gap between the knowledge base and its interpretation. Much literature is available for the understanding of this domain but it is difficult to find something that follows a hands-on approach to the technology. In this chapter, readers will get an introduction of Internet of Things with a special focus on **Bluetooth Low Energy** (**BLE**). There is no problem justifying the fact that the most important technology for the Internet of Things is Bluetooth Low Energy as it is widely available throughout the world and almost every cell phone user keeps this technology in his pocket. The chapter will then go beyond Bluetooth Low Energy and will discuss many other technologies available for the Internet of Things.

In this chapter we'll explore the following topics:

- Introduction to Internet of Things
- Current statistics about IoT and how we are living in a world which is going towards M2M communication
- Technologies in IoT (Bluetooth Low Energy, Bluetooth beacons, Bluetooth Mesh and Wireless Gateways and so on)
- Typical examples of IoT devices (catering wearables, sports gadgets and autonomous vehicles and so on)
- Bluetooth introduction and versioning
- Restructuring of Bluetooth architecture in version 4.0 (BLE) explaining its protocol stack
- Technical terminologies used in BLE and BLE beacons
- Understanding of BLE security mechanism with special focus on Bluetooth pairing, bonding and the key exchange to cover the encryption, privacy, and user data integrity

Internet of Things

The Internet is a system of interconnected devices which uses a full stack of protocols over a number of layers. In early 1960, the first packet-switched network ARPANET was introduced by the United States **Department of Defense** (**DOD**) which used a variety of protocols. Later, with the invention of TCP/IP protocols the possibilities were infinite. Many standards were evolved over time to facilitate the communication between devices over a network. Application layer protocols, routing layer protocols, access layer protocols, and physical layer protocols were designed to successfully transfer the Internet packets from the source address to the destination address. Security risks were also taken care of during this process and now we live in the world where the *Internet* is an essential part of our lives.

The world had progressed quite a far from ARPANET and the scientific communities had realized that the need of connecting more and more devices was inevitable. Thus came the need for more Internet addresses. **Internet Protocol version 6** (**IPv6**) was developed to give support to an almost infinite number of devices. It uses 128 bit address, allowing 2^{128} (3.4 e38) devices to successfully transmit packets over the Internet. With this powerful addressing mechanism, it was now possible to think beyond the traditional communication over the Internet. The availability of more addresses opened the way to connect more and more devices. Although, there are other limitations in expanding the number of connected devices, addressing scheme opened up significant ways.

Modern day IoT

The idea of a modern day Internet of Things is not significantly old. In 2013, the perception of the Internet of Things evolved. The reasons being the merger of wireless technologies, increase the range of wireless communication and significant advancement in embedded technology. It was now possible to connect devices, buildings, light bulbs and theoretically any device which has a power source and can be connected wirelessly. The combination of electronics, software, and network connectivity has already shown enough marvels in the computer industry in the last century and Internet of Things is no different.

Internet of Things is a network of connected devices that are aware of their surrounding. Those devices are constantly or eventually transferring data to its neighboring devices in order to fulfill certain responsibility. These devices can be automobiles, sensors, lights, solar panels, refrigerators, heart monitoring implants or any day-to-day device. These things have their dedicated software's and electronics to support the wireless connectivity. It also implements the protocol stack and the application level programming to achieve the required functionality:

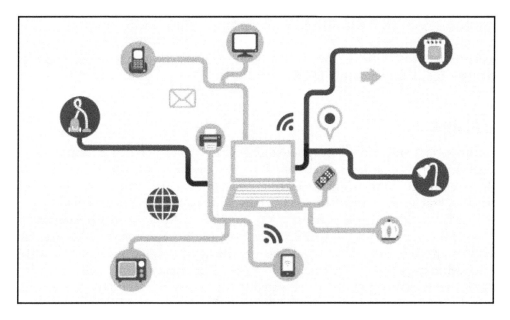

An illustration of connected devices in the Internet of Things

Real life examples of the Internet of Things

Internet of Things is fascinatingly spread in our surroundings and the best way to check it is to go to a shopping mall and turn on your Bluetooth. The devices you will see is merely a drop in the bucket of the Internet of Things. Cars, watches, printers, jackets, cameras, light bulbs, street lights, and other devices that were too simple before are now connected and continuously transferring data. It is to keep in mind that this progress in the Internet of Things is only three years old and it is not improbable to expect that the adoption rate of this technology will be something that we have never seen before.

The last decade tells us that the increase in the Internet users was exponential where it reached the first billion in 2005, the second in 2010 and the third in 2014. Currently, there are 3.4 billion Internet users in the world. Although this trend looks unrealistic, the adoption rate of the Internet of Things is even more excessive. The reports say that by 2020, there will be 50 billion connected devices in the world and 90% of the vehicles will be connected to the Internet. This expansion will bring $19 trillion in profits by the same year. By the end of this year, wearables will become a $6 billion market with 171 million devices sold.

As the section suggests, we will discuss different kinds of IoT devices available in the market today. The section will not cover them all, but to an extent where the reader will get an idea about the possibilities in future. The reader will also be able to define and identify the potential candidates for future IoT devices.

Wearables

The most important and widely recognized form of Internet of Things is wearables. In the traditional definition, wearables can be any item that can be worn. The wearables technology can range from fashion accessories to smart watches. Apple Watch is a prime example of wearables. It contains fitness tracking and health-oriented sensors/apps which work with iOS and other Apple products. A competitor of Apple Watch is Samsung Gear S2 which provides compatibility with Android devices and fitness sensors. Likewise, there are many other manufacturers who are building smart watches, including Motorola, Pebble, Sony, Huawei, Asus, LG and Tag Heuer. The reason that makes them a part of the Internet of Things is that they are more than just watches. It can now transfer data, talk to your phone, read your heart rate and connect directly to Wi-Fi. For example, a watch can now keep track of your steps and transfer this information to the cell phone:

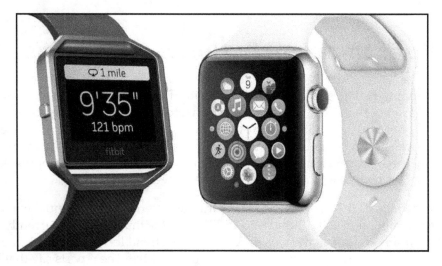

Fitbit Blaze and Apple Watch

The fitness tracker

The fitness tracker is another important example of the Internet of Things where the physical activities of the athlete are monitored and maintained. Fitness wearables are not confined to the bands, there are smart shirts that monitor the fitness goals and progress of the athlete. We will discuss two examples of fitness trackers in this section. Fitbit and Athos smart apparel. The Fitbit Blaze is a new product from the company which resembles a smart watch. Although it can be categorized in the smart watch, the company philosophy makes it a fitness-first watch. It provides step tracking, sleep monitoring, and 24/7 heart rate monitoring. Some of Fitbit's competitors like Garmin Vivoactive watch provides a built-in GPS too. Athos apparel is another example of fitness wearable which provides heart rate and EMG sensors. Unlike fitness tracker, their sensors are spread across the apparel.

 The theoretical definition of wearables may include Augmented and Virtual Reality headsets and Bluetooth earphones/headphones in the list.

Smart home devices

The evolution of the Internet of Things is transforming the way we live our daily lives. People have already started using wearables and many other Internet of Things devices. The next big thing in the field of the Internet of Things is the Smart Home. Home Automation or simply Smart Homes is a concept when we extend our home by including automated controls to the things like heating, ventilation, lighting, air-conditioning, and security. This concept is fully supported by the Internet of Things which demands the connection of devices in an environment. Although the concept of smart home came on the surface in the 1990s, it hardly got any significant popularity in the masses. In the last decade, many smart home devices came into the market by major technology companies.

Amazon Echo

One of the important development in the world of home automation was the launch of Amazon Echo in late 2014. Amazon Echo is a voice-enabled device that performs tasks just by recognizing your voice. The device responds to the name *Alexa*, a keyword that can be used to wake up the device and perform tasks. This keyword can be used followed by a command to perform specific tasks. Some basic commands that can be used to fulfill home automation tasks are:

- Alexa, play some Adele.
- Alexa, play playlist XYZ.

- Alexa, turn the bedroom lights on (Bluetooth enabled lights bulbs (for example Philips Hue) should be present in order to fulfill this command).
- Alexa, turn the heat up to 80 (a connected thermostat should be present to execute this command).
- Alexa, what is the weather?
- Alexa, what is my commute?
- Alexa, play audiobook a Game of Thrones.
- Alexa, Wikipedia Packt Publishing.
- Alexa, how many teaspoons are in one cup?
- Alexa, set a timer for 10 minutes.

With these voice commands, Alexa is fully operable:

Amazon Echo, Amazon Tap and Amazon Dot (from left to right)

Amazon Echo's main connectivity is through Bluetooth and Wi-Fi. It uses Internet network to run the commands. On the other hand, it uses Bluetooth to connect to other devices in the home. For example, the connectivity to Philips Hue and Thermostat is through Bluetooth.

In Google IO 2016, Google announced a smart home device that will use Google as a backbone to perform various tasks. The device will challenge the power of Alexa on the commercial level. This will be a significant step for Google in the smart home market.

 Amazon also launched Amazon Dot and Amazon Tap. Amazon Dot is a smaller version of Echo which does not have speakers. External speakers can be connected to the Dot in order to get full access to Alexa. Amazon Tap is a more affordable, cheaper and wireless version of Amazon Echo.

Wireless bulbs

Philips Hue Wireless Bulb is another example of smart home devices. It is a Bluetooth connected light bulb that give full control to the user through his cellphone. The bulbs can change millions of colors and can be controlled remotely through *away from home* feature. The lights are smart enough to sync with the music:

Illustration of controlling Philips Hue Bulbs with smartphones

Smart refrigerators

Home automation is incomplete without kitchen electronics and Samsung stepped into this race. Family Hub Refrigerator is a smart fridge that let you access the Internet and runs many applications. It is also categorized in the Internet of Things devices as it is fully connected to the Internet and provides various controls to the users:

Samsung Family Hub Refrigerator with touch controls

Television and online media

Internet of Things marked its victory in the Media Electronics as well. An example of the Internet of Things in this domain is the Google Chromecast.

Google Chromecast

It is a digital media player that converts a normal television into a Smart TV. It is designed as a small dongle that connects through HDMI and uses WLAN to wirelessly connect to the cell phone. It lets you cast YouTube, Netflix, Google Photos and many other applications through your cell phone. Google also launched an audio version of the Chromecast that lets you cast your music to any wired speaker.

The advantage of this technology is that it enables a non-smart electronics device to become smart by connecting it remotely to the cell-phone. Google Chromecast was one of its kind portable device to perform a plug-and-play functionality to any television and audio device. Google Chromecast 2nd Generation runs 1.2 GHz ARM-Cortez A7 processor with 512 MB of DDR3L RAM.

 For a complete list of the applications supported by Google Chromecast, follow the link:
`https://en.wikipedia.org/wiki/List_of_apps_with_Google_Cast_supp`
`ort.`

In the following topic, we will discuss how the industry giant Apple is running the best home entertainment smart device:

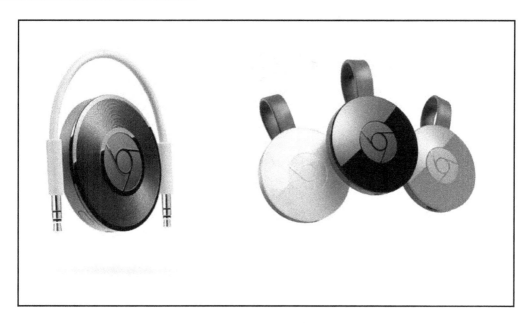

Google Chromecast Audio and Google Chromecast (2nd Generation) from left to right.

Apple TV

While Chromecast provides portability and ease to connect to Android devices, Apple TV runs tvOS (based on iOS) and provides a solid base of the operating system. There are many TV services that provide their services through Apple TV and there is no scarcity of supported applications. It supports Bluetooth, HDMI, Wi-Fi and USB-Type C connectivity. The 4th Generation of Apple TV features an A8 processor with 2GB RAM and up to 64GB of internal storage.

For a complete list of application supported by Apple TV 4th Generation, follow the link
`https://en.wikipedia.org/wiki/Apple_TV#4th_generation_3`.

Moving forward, we will discuss how the Internet of Things made its way in the automotive industry with the help of incredible software stack and smart sensors:

Apple TV 4th Generation

Automotive industry

The discussion on the Internet of Things is incomplete without talking about its impact on the automotive industry. Tesla Motors, a silicon-valley based company was built on the philosophy of smart vehicles. Named after the world's greatest scientist Nikola Tesla, the company sells electric cars, electric vehicle powertrain, and batteries. Tesla Roadster was the first fully electric sports car, launched in 2008. This model was followed by Tesla Model S:

Tesla autonomous driving with GPS navigation on 17" touch screen display

Beginning September 2014, all Model S were equipped with a camera mount on the windshield, forward facing RADAR, and ultrasonic acoustic location sensors giving the car an ability to sense 360-degree buffer. At that time, Model S had the ability to recognize obstacles and road signs. This well-equipped car was given the capability of auto-pilot in October 2015 in the software v7.0 which allows hands-free driving. The update on January (software v7.1) contained "summon" feature that allows the car to park itself automatically without the driver. Another big thing to know here is that the software updates for Tesla are over-the-air unlike some of its competitor. The autonomous driving feature ignites a mixed reaction from the community where some users were excited, others were doubtful about the safety. The update resulted in this car to become a true Internet of Things marvel as it satisfies the requirements perfectly.

The car is now fully aware of its surroundings and can make decisions to navigate safely on the road. Industry experts showed concerns regarding the autonomous driving but Tesla CEO and Google's director of self-driving cars showed their confidence in the technology:

Tesla Model S and Model X (from left to right)

 Tesla Motors has two models after Model S, Model X and Model 3. While Model X is an SUV style vehicle, Model 3 is the most affordable model.

Technologies in the Internet of Things

The Internet of Things is a network of smart devices who are aware of their surroundings. These devices achieve this awareness by constantly or eventually send/receive data with the connected devices. The user, on the other hand, can get access to these devices remotely. In some cases, these IoT devices can let the user perform things autonomously. For example, Tesla cars let the user drive it automatically and Robotic Vacuum Cleaner cleans the house without any master. These functionalities are only possible when these devices are connected and constantly learning about their surroundings. The connectivity is an integral part of these smart devices.

History tells us that the evolution of the Internet was not an easy task. Many standards came and went by before Internet started to converge towards a single most widely used standard. For example, there use to be many networks like ARPANET, UUCP, CYCLADES, NPL and many others before they started to converge under TCP/IP layer model which is now globally used as a standard. It was not before 1984 when CERN began the operation and implementation of TCP/IP as its basic computer networking scheme. The penetration of this new standard was seen in Asia when South Korea adopted TCP/IP communication model in 1982. Australia, on the other hand, was hesitant to adopt this new standard but later in 1989 Australia managed to get rid of their standards before forming AARNet (Australian Academic and Research Network) which provided a dedicated IPv4 based network throughout the country.

The devices in the Internet of Things are connected through one way or another. Sometimes a device can contain multiple modes of communication in order to perform multiple tasks. For example, a smart watch contains Bluetooth to communicate with the cell phone and Wi-Fi to talk directly to the Internet. Similarly, 2nd Generation Chromecast provides Wi-Fi for local and the Internet connectivity and uses HDMI to connect the Television. In IoT, the emphasis is more towards the wireless standards to ensure portability. That is the reason why many experts consider the connectivity of smart devices to smartphones really important. Thus, special emphasis on the wireless technologies will be seen in this book.

Wireless Local Area Network (WLAN)

Wireless Local Area Network or simply WLAN is a network that connects two or more devices wirelessly. A full TCP/IP protocol stack is followed in order to achieve end-to-end communication. This technology is based on IEEE 802.11 standard in which the communication between devices is achieved by a centralized hub/router. The devices are commonly referred to as nodes and the centralized router is referred to as wireless access point. In a usual situation, the wireless access point has a fixed location where the nodes can freely roam around within the coverage area of the access point. The connected devices can be tablet computers, laptops, desktops, cell phones or any IoT device (such as a smartwatch or Google Chromecast). Residentially, a typical access point covers an apartment but in commercial situations multiple access points can be bind to one SSID. If a node travels from Router 1's coverage area to Router 2's coverage area, they will perform a seamless handover between each other. Modern day WLAN are advertised under the Wi-Fi brand name.

Following mentioned diagram shows how Wi-Fi router communicates to its nodes. It is a typical working model of a WLAN:

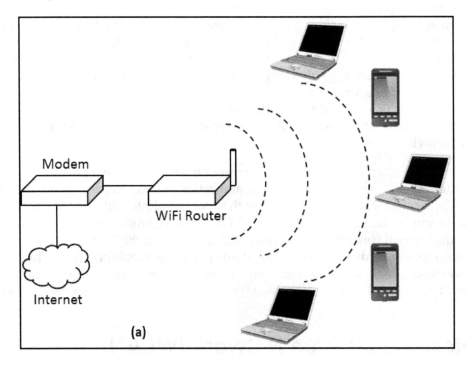

(a)

Wireless Local Area Network architecture

This technology is proven to be very effective in modern day Internet of Things because the architecture is simple and can easily be implemented on a device with limited capabilities. Moreover, IPv6 gives us approximately 3 x 10^38 addresses which are more than enough to give identity to every atom in the world, let alone smart devices.

 WLAN is sold under the name Wi-Fi and can work on five different spectrums. 5.9 GHz, 5 GHz, 4.9 GHz, 3.6 GHz and 2.4 GHz. Each frequency range is further divided into many channels and countries impose their radio regulations independently.

Wireless Local Area Network is an important technology in the Internet of Things and devices like smart watches, media players and autonomous vehicles often use this technology to transfer information. Sometimes, this technology is used to give a control to the smart devices over a long distance.

Wireless Ad-hoc Networks

Wireless Ad-hoc Networks lies under the umbrella of IEEE 802.11 standard as well but unlike WLAN, they don't have a centralized hub/router. The devices are responsible for routing the packets across the networks. **Wireless Mesh Network (WMN)** is a sibling of Ad-hoc Networks with the only difference of traffic abnormalities and mobility. Ad-hoc networks are more dynamics as compare to WMN and they have more traffic irregularity than WMN.

In Ad-hoc networks, the device acts as a router to transmit traffic between peers if the destination device is not directly connected to the sender device. This formation of devices is reliable because each device is connected to another device in a mesh, giving alternate routes to any possible incoming packet. These networks are also self-organized, so if a device leaves the network, the network will reconfigure itself for new potential routes for forwarding packets. Due to their architecture, they are widely used in small and large networks. They run on TCP/IP based model and can easily be implemented in any lite-weight smart device.

Since they are connected to one another directly, they are also called peer-to-peer networks. The architecture of Ad-hoc networks is different than WLAN which brings advantage and disadvantage at the same time. Ad-hoc networks are fast and don't contain a single point of failure like a router, but they are more prone to network attacks. Ad-hoc networks are easily scalable networks where devices can come and go at their will. On the other hand, those devices are responsible for routing the packets, which make them weak for Man in the Middle attacks.

Wireless Ad-hoc Networks are further classified into five types:

- **Mobile Ad-hoc Networks (MANETs)**
- **Smartphone Ad-hoc Networks (SPANs)**
- **Vehicular Ad-hoc Networks (VANETs)**
- **Internet-based Mobile Ad-hoc Networks (iMANETs)**
- **Military Mobile Ad-hoc Networks**

Further information on Ad-hoc networks can be found on `https://en.wi`
`kipedia.org/wiki/Wireless_ad_hoc_network`.

The inclusion of Ad-hoc networks in the smart device makes them vulnerable because they normally contain personal and sensitive information. This brings a threat to the Internet of Things. On the other hand, Ad-hoc is almost the perfect topology for IoT devices as they are fast, scalable, easily maintainable and cheap.

To understand how a mesh is connected, consult the following figure:

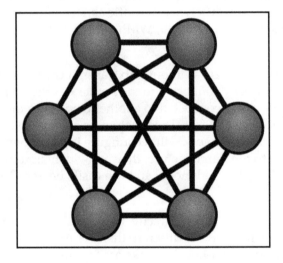

An illustration of a fully connected Ad-hoc Network

ZigBee

ZigBee is a wireless mesh network built on IEEE 802.15.4 standard. With the advancement in sensors and low-powered smart devices, there was a need for a technology that can connect them wirelessly in the form of a mesh. The ZigBee technology was introduced which works on a Personal Area Network and works on low power. Internet of Things searches for the technology with low power consumption because it is built with the devices with limited power capabilities. For that purpose, ZigBee is considered as a technology with a seamless connection flexibility, ideal power consumption, and reliability.

ZigBee gives a 10-100m line of sight coverage for connectivity. Typically, it works on a 2.4 GHz band (ISM) but can be found on other frequency like 784 MHz, 868 MHz, and 915 MHz. With the wide variety in the spectrum, the data rates in ZigBee varies from 20 Kbit/s to 300 Kbit/s. The ZigBee module can be of 0.5 inches which are easily implementable in any Internet of Things device. One of the attractive parts of ZigBee is that it has out of the box capabilities of wireless mesh, star and tree architecture:

ZigBee Logo

A significant disadvantage of ZigBee lies in its adoption in computing devices. Typically, a smartphone or a laptop does not come with this module. Computers and cellular phones are the basic connection points for IoT devices. For example, if a person has ZigBee-based smart-home device, he will not be able to control it with his cellular phone or laptop without connecting additional hardware. This is a very crucial disadvantage as IoT devices mostly, if not always, rely on a smartphone-based remote control. Philips Hue (a smart bulb) can not be remotely controlled by a smartphone if its major technology of communication is ZigBee.

This disadvantage leaves a question mark on the technology but the other side can not be neglected as well. ZigBee is the only lite-weighted-low-power-consumption technology available in the market which supports mesh communication out of the box. Companies like smart plugs and smart energy interface use ZigBee as their main mode of communication. Let's take the example of Philips Hue Wireless dimming kit. It comes with a separate controller which talks to the smart bulb without any need of the smartphone. In this scenario, ZigBee can be the most practical solution.

A demonstration of the Philips Hue smart bulb can be seen here:

Phillips Hue wireless dimming kit with ZigBee technology

Bluetooth Low Energy

Started by Dr. Nils Rydbeck, Bluetooth was first conceptualized in 1989 and later built by Ericsson in 1994. The name Bluetooth was given after the tenth-century king of Denmark, Herald Bluetooth. The king united Danish tribes and introduced them to Christianity. This name was given to the technology by Jim Kardach in 1997. The name was adopted because the core concept of Bluetooth was to build a system that can connect a phone to the computer. The idea of Bluetooth was supported by Nokia and Intel who were thinking to migrate towards a universally interoperable wireless technology. Later, companies including Ericsson, Intel, Toshiba, Nokia and IBM came together in Lund, Sweden, in 1996 to agree on the foundation of **Special Interest Group** (**SIG**). This group is responsible for designing and maintaining the Bluetooth technology. Nowadays, the group is called *Bluetooth SIG*.

King Herald Blatand was given the nickname Bluetooth because of his habit of eating blueberries.

Bluetooth technology was invented to connect mobile devices to computers over a short distance. It was standardized by IEEE 802.15 working group which describes Wireless **Personal Area Network (WPAN)** standards. IEEE 802.15.1 defines physical and Media Access Control (MAC) layer specifications for short distance wireless connectivity for Bluetooth. Version 1.0 was developed by the SIG but resulted in interoperability problems which were later fixed in v1.1. The technology was never to be used as a replacement of any wireless Internet technology like Wi-Fi. It was to only use for the limited range and with a limited amount of data. Even though Bluetooth today can transfer a huge amount of data, it is still being used for short distance communication only.

Bluetooth versions

From 1999 till 2009, many versions of Bluetooth were released. Solving many problems from the initial release while increase the speed and the connection quality. It was not until the version 3.0 when Bluetooth could achieve a speed of 24 Mbit/s. During this evolution, a continuous change in the protocol stack can be observed to solve various issues in the technology. For example, in Bluetooth v3.0, **Logical Link Control and Adaption Protocol (L2CAP)** was introduced to multiplex multiple logical connections between various devices using different protocols. **Service Discovery Protocol (SDP)** was also introduced to connect to the Bluetooth devices like headsets etc. During this period of time, the member companies of Bluetooth SIG rose to 12,000 and by 2010, it surpassed a total of 13,000 companies.

The next step in the history of Bluetooth was a significant one. The SIG decided to take a step back and rethink the protocol stack being used in Classic Bluetooth. Bluetooth Core Specification 4.0 or simply Bluetooth Low Energy came into being in 2011. It was previously known as Wibree by Nokia which replaced the entire protocol stack of the previous technology and concentrated on the speed and the simple architecture. 2011 was the same year when SIG surpassed 15,000 companies, Apple and Nordic Semiconductor joined SIG Board of Directors, Apple released first two computers with Bluetooth 4.0 support and Microsoft announced Windows 8 phone with Bluetooth 4.0 Core Specifications.

Unlike classic Bluetooth, Bluetooth Low Energy focused on the low power consumption. That was a very important step from the SIG as it was realized that in future, the technology will not be just fast but less power hungry as well:

Bluetooth SIG official Logo

Bluetooth v4.1 came as a software update rather a hardware update. The update focused on the usability and implemented Bluetooth Core Specification Addenda. Previously, the Bluetooth radio and LTE radio didn't coexist and resulted in interference. The issue was causing performance issue and fast battery drainage. Bluetooth 4.1 enabled both these technologies to co-exist in one device with high data rates. BLE v4.1 was the first version which enabled any device to become a server and client. On the other hand, the smart connectivity was introduced which let devices manage their power better by enabling manufacturers to define their own reconnection time intervals. The version of Bluetooth was commercially available in early 2013. In the same year, Google also announced the native BLE support to Android.

With power efficiency and server/client flexibility, Bluetooth Low Energy officially stepped into the Internet of Things.

The evolution in the Bluetooth leads us to Bluetooth v4.2. In December 2014, Bluetooth SIG announced the next version of Bluetooth with some special features supporting IoT. Bluetooth Low Energy v4.2 provided flexible Internet connectivity options to achieve power efficiency. IPv6 support was given to the standard with an ability to make Bluetooth Smart Internet Gateway (GATT) architecture. The concept will be discussed in detail later in the chapter.

In Bluetooth 4.2, LE privacy and LE Secure Connections were also introduced by Bluetooth SIG. Two-factor authentication via BLE was introduced to provide better security. The new version also claimed that it is not 2.5 times faster with 10-fold extension in data length. The over-the-air firmware update was also introduced in this version which provides manufacturers to remotely update the firmware without even touching the device. Bluetooth 4.2 is backward compatible with version 4.0 and 4.1 which were also Bluetooth Low Energy.

iOS, Android, Windows Phone, BlackBerry, OS X, Linux and Windows 8 natively support Bluetooth Low Energy.

Bluetooth Low Energy Key Concepts

Bluetooth Low Energy is a standard with millions of devices running it. Discussing Bluetooth Core Specification in full-length is almost an impossible task as it has many technicalities. So the key concepts that will help you develop the Bluetooth Low Energy applications will be discussed in the book.

The layer stack of Bluetooth Low Energy corresponds to the OSI layer. In this book, we will not discuss the lower level layers (that is, from PHY to L2CAP) as they are out of the scope. The main concepts covered in this book will be Attribute Protocol, **Generic Attribute Profile** (**GATT**), **Generic Access Profile** (**GAP**) and the top layer Bluetooth Application. The book will give practical examples of how to write the Bluetooth Application based on the GATT server/client architecture.

The protocol stack of the Bluetooth Low Energy can be seen in the following figure:

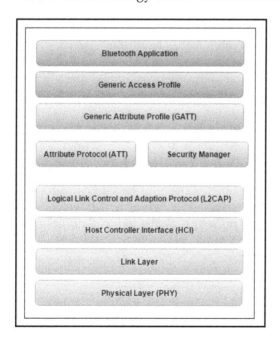

Bluetooth Layered Model

Attribute Protocol (ATT) and Generic Attribute Profile (GATT)

Bluetooth Low Energy brought two core specifications and every Low Energy profile is supposed to use them. Attribute Protocol and Generic Attribute Profile.

Attribute Protocol is a low-level layer that defines how to transfer data. It identifies the device discovery, reading and writing attributes on a fellow device. On the other hand, Generic Attribute Profile is built on the top of ATT to give high-level services to the manufacturer implementing LE. These services are basically used to manage the data transfer process in a more systematic way. For example, GATT defines if a device's role is going to be Server or Client.

An interesting thing about ATT and GATT is that they are not transport-layer specifications, that means that they can be implemented on BR/EDR or LE. GATT is a mandatory entity in LE and used to discover services and characteristics. The GATT server listens to an ATT requests and confirmations sent by GATT client. GATT server stores, process and transfer the data to the client. Another role of the GATT is that it defines the data arrangement on the server side so that the client can read it accordingly. The data transfer between GATT server and GATT client is called an "Attribute". An attribute is uniquely identified by a **Universally Unique Identifier (UUID)** which is 128 bits long string ID.

The Bluetooth system consists of four base layers. Radio (Physical layer), Baseband, Link Layer, and L2CAP.
More information about the working on these layers can be found in the core specification document: `https://www.bluetooth.com/specificatio ns/adopted-specifications`.

GATT Server and GATT Client

Any device that wants to transfer data over Bluetooth can adopt a role based on the requirement. A device can act as a GATT Server which welcomes requests from the client, process data accordingly and returns back the value. GATT Server is responsible for processing and making data attribute available to the client.

GATT client, on the other hand, initiates the request to the server and receive the responses. A GATT client needs to perform a service discovery process in order to know the server's attributes. Once it is done identifying the services, it can read or write data attribute.

Universally Unique Identifiers (UUIDs)

A UUID is a unique identifier that is guaranteed to be globally unique all the time. It is a 128-bit identifier out of which some pre-allocated values are used for registration purposes as described by the Bluetooth SIG. One of the very first pre-allocated values for UUID is known as Bluetooth Base UUID and has a value of 00000000-0000-1000-8000- 00805F9B34FB. Pre-allocated values in the UUID often has 16-bits or 32-bits aliases that can be used to generate your own UUIDs. The formula for generating a 128-bits UUID with pre-allocated alias is:

```
128-bit-UUID = 16-bit-value * 2 ^ 96 + Bluetooth-Base-UUID
128-bit-UUID = 32-bit-value * 2 ^ 96 + Bluetooth-Base-UUID
```

Bluetooth SIG provides UUIDS for services and profiles and they are made for a specific purpose. If it doesn't specify a particular need, a new UUID can be created. Using the Bluetooth-Base-UUID is recommended by the SIG for this purpose.

GATT-based Bluetooth Profile Hierarchy

Generic Attribute Profile describes the structure following which a profile exchange data from one device to another. The basic elements of a profile are services and characteristics. Typically, a profile is made in order to achieve the desired task. Services and characteristics are the main components of a profile which can be described as the subtasks.

The division of the GATT-based hierarchy is given here:

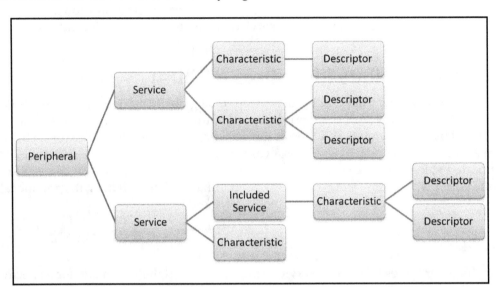

GATT based profile model with services, characteristics, and descriptors.

Service

A service is a collection of characteristics designed to fulfil a particular task. It can also reference another sub-service which then comprises of characteristics. For example, a profile can have a Heart Rate Monitor service which is to expose heart rate for fitness purposes. This service can have Heart Rate Measurement, Body Sensor Location, and Heart Rate Control Point characteristics which are the subtasks of the Heart Rate Monitor service.

Characteristic

A characteristic contains a value, properties, and configuration information. Characteristic is the combination of characteristic properties, characteristic declaration, characteristic value, and a descriptor. Characteristic properties define the level of access that characteristic grants. For example, if a characteristic is read-only it will be defined in the properties. The characteristic declaration can be seen as a characteristic overview where the name, description, type and the requirements of the characteristic is mentioned. Descriptors are used to define a characteristic value.

The following diagram shows a view of the BLE characteristics:

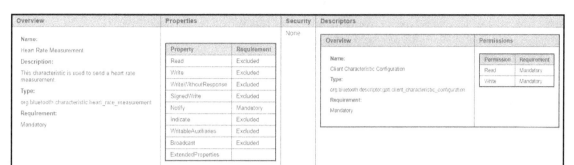

Heart Rate Measurement characteristic with overview, properties, security and descriptor

The example in the figure is describing a **Heart Rate Measurement** characteristic. The overview describes the functionality of the characteristics and the requirement level. The requirement tells if it is mandatory or optional for the service. The properties show the access-level of this characteristic. In this particular scenario, the characteristic is only used to notify. The descriptor identifies the value or permits configurations of the GATT server with respect to the characteristic value.

Generic Access Profile (GAP)

Bluetooth Low Energy specifies a generic profile that enables BLE devices to communicate with each other. It can be understood as a base profile which is the same throughout the BLE devices. It defines basic requirements of a Bluetooth device. This generic profile is used to tie all the layers together and identifies all the layers from PHY to L2CAP. It helps to universally maintain all the layers of Bluetooth Low Energy.

Generic Access Profile defines security by implementing cryptographic algorithms that enable devices to securely transfer data attributes. This functionality is achieved by Security Manager which is implemented on the protocol stack. As GAP is universally spread over all the layers, it can maintain and dictate the standard to all the layers. The **Advanced Encryption Standard** (**AES**) is used for secure key exchange between the peers to start a communication. GAP sets the procedures for security by giving the trust to the peers to carry sensitive data across the Bluetooth connection.

Generic Access Profile (GAP) also defines the roles used in the Bluetooth Low Energy. These roles are Broadcaster, Observer, Peripheral, and Central. These roles are really important to understand as the foundation of BLE is based on it. Based on the implementation, a BLE device can operate under one role at a time. This multi-role scenario is only possible if the underlying controllers support those roles.

Broadcaster

Broadcaster is a role to the device that wants to periodically or constantly broadcast the information. A Bluetooth beacon is a good example of this when all it needs to do is to broadcast information. Theoretically, the role of broadcasting is given to the transmit-only devices, but practically it can be given to any device that transmits and receives. Another important information about the broadcasters it that the data sent is not a result of any connection. Broadcaster always transmits data irrespective of who is listening. These packets are special advertising packets and should not be confused with the connection data packets from GATT Server and Client.

Observer

The role of Observer is to listen to the packets broadcast from the Broadcaster. Typically, they are meant to be for receive-only applications. The observer is able to read the advertised packets and listen to the data coming out of the Broadcaster. Just like Broadcasters, they do not support incoming connections.

Central

The role of Central can be understood as a role of master in a master-slave architecture. It is capable of making a connection to the peer. A central can connect to multiple peers at the same time. A central can be a smartphone/tablet which can initiate a session by connecting to various BLE peer. A smartphone has the capability to connect to the Bluetooth headphones and Bluetooth smartwatch both at the same time. Central is also the originator of the connection request which means that no other role can start the connection request.

Peripheral

Peripheral, on the other hand, resembles a slave in a master-slave architecture. It broadcasts advertisement packets for the central so that it can find it and then accepts the connection request. Even though it advertises the data, it is power and processing efficient.

Bluetooth Low Energy works on 40 channels with 2MHz spacing where the discovery is done by advertising. BLE supports a maximum data rate of about 1Mbps and provides AES-CCM encryption for confidentiality. 50-100 meters is the practical range of BLE with 10mW maximum power output.

Bluetooth Low Energy Security

Bluetooth Low Energy provides five features for its security including pairing, bonding, encryption, authentication and message integrity. These satisfy three pillars of security which are Authentication, Confidentiality, and Authorization. Attacks against improperly secured Bluetooth implementation can provide hackers with unauthorized access to the sensitive information and unauthorized use of the Bluetooth device. The main security features in Bluetooth Low Energy are:

- Pairing is the process of generating shared keys (on both ends) known as **Short Term Key (STK)**
- Bonding is the subsequent process of pairing where they store the STK in order to form a trusted pair
- Device authentication is when two devices identify if they have same keys
- Encryption is for message confidentiality
- Message integrity is to avoid attackers to forge the data

Key Generation

Key generation is a process done while a device wants to pair with another device. The process is independent of any other LE device. The reason to generate a key and pair the device with another device are:

- Securing device identity
- Authentication of the unencrypted data.
- Confidentiality

For key generation purposes the devices exchange required information and then the **Short Term Key** (**STK**) is calculated. The link is then encrypted used AES-CCM cryptography. While the key generation is not performed on the controller, the actual encryption is performed on the controller.

Association Models in Pairing

These models describe the negotiations between the device to generate **Short Term Key** (**STK**). BLE provides four association models:

- Numeric Comparison
- Just Works
- **Out of Bands** (**OOB**)
- Passkey Entry

Numeric Comparison

This model is defined when both BLE devices are designed with a display. This display is used to show a six-digit number and both are capable to verifies those numbers with a simple yes or no. For example, a user is shown a six-digit number 453145 on both screens and then asked if the numbers appeared on both ends are same. If the user answer yes, the pairing is successful. This model provides security from man-in-the-middle attacks and also serves the purpose if the user is connected to the right device.

Just Works

This model is defined for the scenarios where one or more devices do not have a display to show any pin or any digit. In this model, the user is not shown any digits, rather, a numeric comparison happens and then the user is asked to accept the connection. This model is not secure from **man-in-the-middle** (**MITM**) attack , where a person is listening to the channel constantly.

An example of this model is when a cellular phone tries to connect to a headset. At the headset end, there is no display to show the digits. Hence, the numeric comparison happens and the user is just asked to accept the connection.

Out of Band (OOB)

This model is applicable in the scenarios where devices are capable of using another technology like NFC. It is called Out of Band mechanism to discover both devices as well as to exchange cryptographic numbers. The model provides security from MITM attacks. The practical example of this model is NFC-based connection between a cellular phone and a Bluetooth headphone. The user needs to touch the headset in order to pair it with the phone.

Passkey Entry

This model is for the scenarios where one device has input functionality but not the display functionality and other devices have only display functionality. The user is shown a six-digit number 456654 on the device with a display and then asked to enter the number on the other device. If the number entered in the other device is correct, the connection is paired.

 According to Bluetooth SIG, Just Works and Passkey Entry do not provide any passive eavesdropping protection. This is because Secure Simple Pairing uses Elliptic Curve Diffie-Hellman and LE legacy pairing does not.

Signing the Data

Signing the data is a very important concept in terms of message integrity. If the encryption is not available, BLE has the ability to send authenticated data over an unencrypted link between two devices. This is done by signing a data using a **Connection Signature Resolving Key** (**CSRK**). This signature is placed after the **Packet Data Unit** (**PDU**). The receiving device verifies the signature and considers it coming from a trusted source. The signature is made by a **Message Authentication Code** (**MAC**) which can sometime use Hash Functions. This MAC is made upon counter value which protects the replay attack.

Privacy

Bluetooth Low Energy provides a unique feature that diminishes the ability to track an LE device by changing the Bluetooth device address frequently. This is an important privacy feature as it allows an advertiser to hide. These Bluetooth device addresses are temporary and are randomly generated that can only be recognized by bonded device. Whenever a device is bonded to another device for the first time, an **identity key** (**IRK**) is generated which is later used to generate a private address. This address must be resolvable by the other devices in order to keep the privacy.

Bluetooth Low Energy for the Internet of Things

Bluetooth Low Energy was built to fulfill the requirements of the Internet of Things. Bluetooth SIG took a significant step when they decided to revamp the protocol stack of the Bluetooth. Bluetooth Classic was successfully transformed into Bluetooth Low Energy and adopted rapidly in the IoT devices and more importantly, in cellular phones. Bluetooth low energy can be summarized in two basic factors: power efficiency and simple architecture. While ZigBee is commercially successful and has been used for quite a while, Wi-Fi and Bluetooth Low Energy are the only famous and widely spread standards used in consumer electronics. If the judgment criteria are strictly on the basis of the adoption rate, Bluetooth Low Energy surpasses ZigBee as the adoption rate of the ZigBee is almost zero in the smartphones.

Applications of BLE in IoT

In the past five years, the work in the Internet of Things is significant in comparison to the previous times. Many manufacturers and start-ups are focusing on this technology in order to make accessories for the smartphones. The reason manufacturers are more focused towards the BLE as compared to any other technology is that they want to increase the adaptability of their IoT devices. As most of the smartphones exist have Bluetooth technology in it, the adoption rate of these accessories is tremendous as compared to any other standard. People want to use their smartphones to control and interact with things. It would be great if they can control the bedroom light from their smartphone or turn off the stove from their cell phone.

The promity sensing, indoor navigation, fitness tracking and other countless applications have already found their way in the Bluetooth Low Energy. Users do not have to sacrifies the battery life while enjoying the connection with their favorite IoT devices. At this moment, it is not clear if the future of IoT lies in the combination of several wireless standards in one device (for example, combining Wi-Fi and BLE both in a single device) or there should be one stand-alone standard which is the master of everything. It is sensible to predict that there is no significant loss if there are multiple standards in one device as the architecture and technology become cheaper. The companies in consumer electronics will not hesitate to go to multi-standard solutions in the longer run.

Bluetooth World event organized by Bluetooth SIG in Santa Clara, United States

Companies like Google, Apple, and Intel are already backing up the BLE standard. The progress and the agreement of the industry can be seen in the work of the Bluetooth SIG because as of June 2016, SIG announced a new version of BLE as Bluetooth 5. Quadrupling the range and doubling the speed will make Bluetooth Low Energy compete more aggressively in the market. In Bluetooth World 2016, which is an annual event organized by the Special Interest Group, many manufacturers were present to show their products. It showed the diversity how companies are changing the way we think about Bluetooth technology. CSRMesh which is acquired by Qualcomm, showed off their mesh technology built over Bluetooth to take this standard to another level and truly threat ZigBee as it is the only standard that provides a low powered mesh architecture.

Bluetooth Low Energy Beacons

The modes of communication of Bluetooth Low Energy are already discussed in this chapter that is why it is a good point to introduce the Bluetooth Low Energy Beacons. They are the broadcast only devices that advertise certain information to the users around them. This functionality enables a user to receive information from the beacon and in some cases take actions too. These broadcast happens in a limited proximity and can be changed by modifying the properties of the beacons.

Bluetooth Low Energy Beacon use-case

A use-case of Bluetooth Low Energy Beacons is their implementation in the bus stops. Instead of putting the bus timings on the transit board, it would be better if the bus stop broadcasts this information through Bluetooth. Since BLE is a low power consuming technology, it is completely feasible for the user to receive this information from the bus stop. Another use-case can be the boutique sales. Instead of putting up the banners, shops can broadcast the information through Bluetooth beacons and the shoppers around them receive a notification in their cell-phone about the sale.

The battle between Google and Apple can be seen by these logos:

Apple iBeacon and Google Eddystone logos

Communication model of BLE Beacons

Bluetooth Low Energy Beacons are based on the one-way communication model. This is important because it will be inappropriate if someone changes the broadcast data on the beacons. Bluetooth Low Energy Beacons are supported by two of the best companies in the consumer electronics market. Google and Apple have their own protocols made for Bluetooth beacons. Google Eddystone is an Apache 2.0-licensed cross platform profile available for Bluetooth beacons. It works on different frame types like Eddystone-UID, Eddystone-URL, and Eddystrong-TLM. iBeacons, on the other hand, is a protocol developed by Apple in 2013 serving the same purpose. These topics are discussed in a greater detail in `Chapter 4`, *Bluetooth Low Energy Beacons*.

Bluetooth mesh networks

The mesh networking is a proven technology in terms of design and scalability. Information Technology has already seen the advantages of this technology in the form of wireless mesh networks, wireless sensor networks, mobile ad-hoc networks and vehicular ad-hoc networks. In the Bluetooth World 2016, SIG showed their interest in the mesh technology and announced that they are working with Qualcomm to build a first generation mesh technology by next year.

The mesh technology is a significant advantage of ZigBee in the latest times because it is fully developed and fully functional mesh mechanism to support upto 65,000 nodes mesh. Bluetooth, on the other hand, does not support mesh technology. By entering in the mesh networking, Bluetooth Low Energy will become powerful enough to replace ZigBee in the commercial industry. Logical prediction can be made that the adoption of the Bluetooth mesh will be much faster as it will be supported by the cellular phones, tablets, and laptops.

A much detailed overview and a practical hands-on approach on the Bluetooth Mesh will be discussed in `Chapter 6`, *Bluetooth Mesh Technology*.

Summary

In this chapter we spoke about the Internet of Things technology and how it is rooting in our real lives. The introduction of the Internet of Things was given in the form of wearable devices, autonomous vehicles, smart light bulbs, and portable media streaming devices. Internet of Things technologies like wireless local area network, mobile ad-hoc networks and ZigBee was discussed in order to have a better understanding of the available choices in the IoT. The technology discussion led us to the Bluetooth Low Energy, where the concepts and the technical structure of the Bluetooth Low Energy were discussed in great detail. The chapter introduced us to the Bluetooth stack and the key concepts like Attribute Profile, Generic Attribute Profile, GATT Server, GATT Client, UUIDs, Generic Access Profile, Service, Characteristics and the Bluetooth Security. Later we discussed Key Generation mechanism in the Bluetooth Low Energy, pairing, bonding, privacy and signing mechanisms. In the end we concluded with two major IoT platforms Bluetooth Low Energy brings and how Bluetooth beacons and Bluetooth Mesh can be the next big things in the Internet of Things.

In the next chapter we'll discuss different Hardware and Software tools available for Bluetooth Low Energy development.

2

BLE Hardware, Software, and Debugging Tools

Bluetooth Low Energy, a successor to the Classic Bluetooth, is a young technology with huge adoption rate and big market. The technology made itself visible in mid 2010; when big manufacturers like Apple launched their products in the market with Bluetooth Low Energy support. Not only this, but Bluetooth version 4.0 took care of the biggest dilemma of the previous decade. It now consumed less power as compare to its predecessor. Bluetooth **Special Interest Group** (**SIG**), the developers and maintainers of Bluetooth technology, made sure that the technology matched the trend of the market and helped to spread itself in the **Internet of Things** (**IoT**). Chapter 1, *BLE and the Internet of Things*, described the context of Bluetooth Low Energy in the world of the Internet of Things and how it got popular amongst the manufacturer as a cheap short-distance communication technology. This chapter will be the next step in fulfilling the motivation of this book, which is to give readers a complete guide, starting from the very beginning till implementing their own projects.

In this chapter, the book will define the following topics:

- Bluetooth Low Energy tools for development
- Bluetooth Low Energy hardware including Nordic nRF51 development kit (with SoftDevices), Coin Arduino BLE Kit and Adafruit Bluefruit LE development kit
- Bluetooth Low Energy software including Bluetooth Developer Studio by SIG, nRF Connect, nRF Logger, nRF UART 2.0, nRF BLE Joiner, nRF Beacon for Eddystone, Google Beacon tools and physical web

Bluetooth Low Energy hardware

Classic Bluetooth was famous for its short distance connectivity in a cellular phone. This communication was a breakthrough over conventional IR communication as it did not require line-of-sight communication. The result was marvelous. It got famous among the cellular phone users and almost every vendor started to support Classic Bluetooth. The idea of the Bluetooth was simple enough for any lay person. You did not require any technical knowledge in order to send files and pictures to someone else's device. All you need to do is to enter the same passcode on two devices before pairing them.

It is not untrue to say that the success of Classic Bluetooth was marked by cellular industry. Users of Bluetooth did not require any additional Bluetooth hardware, rather their personal phones contained the built-in facility to connect to another Bluetooth device. Before the concept of Bluetooth Low Energy, the main hardware for Bluetooth communication was a cellular phone. This situation changed when Bluetooth Low Energy came into being. The industry realized that the devices can now be powered with the low energy. Users can now own their personal Bluetooth Low Energy device for the cheaper price.

This change in behavior in the thinking caused a boom in the Internet of Things. Users now have a technology that can be put into their appliances and provides free connectivity to their cellular phones. For the first time, Bluetooth came out of the cellular phones and was put separately. It is not to forget that Bluetooth headphones and some other Bluetooth devices existed before the launch of Bluetooth Low Energy but it came at the cost of the high power consumption.

Bluetooth Low Energy hardware is now a separate industry where manufacturers are making it easier for people to deploy BLE in their own devices. Bluetooth Developer Kits are widely available for the use today. This chapter will talk about some famous Bluetooth Low Energy hardware available in the market today.

Development kits

Development kit is typically a set of tools provided for the ease of the developer. It helps to build hardware and software applications and reduces significant amount of time.

Software developers use development kits all the time to save their time and implement pre-built APIs in their application. For example, Android developers use Android **Software Development Kit** (**SDK**) to create Android applications. This SDK enables developer to implement **Application Program Interface** (**API**).

Hardware development kit normally comprises of a hardware component such as a programmable circuit board and a set of API to program it. For example, a Bluetooth development kit normally comes with a ready-to-program circuit board, a software API and its documentation. In this chapter we will discuss some Bluetooth Low Energy development kits and what they bring on the table for the developers.

Nordic semiconductor nRF51 development kit

Nordic Semiconductor is a semiconductor company that specializes in low power wireless **systems-of-chip** (**SoC**) with general connectivity over 2.4 GHz ISM band. Their focus is to provide low power consumption circuits for consumer electronics. The company provides solutions like sports equipment, Bluetooth SoC, wireless medical equipment and wireless audio applications.

Nordic Semiconductor nRF51 DK is a single board development platform to develop Bluetooth Low Energy applications. It also works for 2.4 GHz applications using nRF51 Series SoC (also provided by Nordic). It is based on 16 MHz ARM Cortex-M0 CPU and very useful for prototyping and building Bluetooth Low Energy applications. The Kit is easily connected with Master Emulator Firmware and the Master Control Panel Software to sniff BLE packets and the communicate with other Bluetooth Low Energy devices. The kit consists of GPIO pins with four programmable buttons and four LEDs. The nRF51 DK comes with 256 KB flash, 32 KB RAM and a built-in power supply. It also comes with a micro-USB port to connect the DK with the computer.

The Nordic board can be found here:

nRF51 DK with two GPIO connected boards

The development kit is highly capable of taking advantage of Bluetooth Low Energy flexibility. It can perform **Over-The-Air Device Firmware Updates (OTA - DFU)**. That means that the we can update the firmware on the kit just by using our BLE enabled cellular phones. It is also capable of running any BLE application like beacons, smart watches, fitness tracker, smart home and RC toys.

Nordic's nRF51 development kit comes with full software support. Nordic Semiconductors introduces their protocol stack by calling it SoftDevices. These SoftDevices act as a middleman between Nordic's application layer and physical layer. SoftDevice contains the implementation of GATT, GAP, L2Cap and security mechanisms. (You can refer the previous chapter for an introduction of these terms).

This protocol stack can be downloaded from Nordic's website and can easily be implemented on nRF51 system-on-chips. There are many SoftDevices available in the Nordic's website and can be implemented on SoC in line with requirements:

- **S110**: Bluetooth Smart Peripheral Stack
- **S120**: Bluetooth Smart 8-link Central Stack
- **S130**: Bluetooth Smart Concurrent central/peripheral/broadcaster/observer Stack
- **S210**: ANT/ANT+8 link Stack
- **S310**: Bluetooth Smart Peripheral/ANT 8-link Stack

The Nordic logos for these SoftDevices are:

SoftDevices logos on the Nordic Semiconductor's website

S110 SoftDevice

S110 is a Bluetooth Low Energy protocol stack solution with Peripheral/Broadcaster functionality. It provides BLE controller and host functionality with flexible API. This enables developers to build Bluetooth Low Energy SoC applications with minimum efforts. Some of the key features of this SoftDevices are:

- Bluetooth 4.2 enabled single-mode protocol stack
- L2CAP, ATT, GATT (client and server) and GAP
- SMP support with MITM and OOB pairing
- Bluetooth examples to implement custom applications
- ARM Cortex-M0 project configuration for application development
- Master Boot Record for **Over-The-Air (OTA)** DFU

S120 SoftDevice

S120 is a Bluetooth Low Energy protocol stack solution with up to eight Central role communication. Like S110, it provides a full and flexible API to build BLE SoC applications. Some of the key features of this SoftDevices are:

- Bluetooth 4.2 enabled single-mode protocol stack
- Thread-safe based API
- Link Layer
- L2CAP, ATT, GATT (client and server) and GAP
- Central/Observer role with up to 8 simultaneous connection
- SM Including MITM and OOB pairing
- Bluetooth examples to implement custom applications
- ARM Cortex-M0 project configuration for application development

S130 SoftDevice

S130 is a Bluetooth Low Energy multi-link protocol stack solution with four role connections. It can perform central, peripheral, broadcaster and observer connections. BLE controller and host are provided with flexible API to build BLE SoC applications. The key features of S130 are:

- Bluetooth 4.2 enabled single-mode protocol stack
- Thread-safe based API
- Link Layer
- L2CAP, ATT, GATT (client and server) and GAP
- Central, Peripheral, Observer, Broadcaster roles
- SM Including MITM and OOB pairing
- Bluetooth examples to implement custom applications
- ARM Cortex-M0 project configuration for application development

 ANT is an ultra-low power, short distance wireless technology that works on 2.4 GHz ISM band. S210 is for ANT based networks and is not discussed in this book. However, S310 is a hybrid between ANT and BLE. For more information about these SoftDevices, you can consult Nordic website or ANT forum `https://www.thisisant.com/forum/viewthread/4603`

Find the Nordic board diagram with their technical detail as shown here:

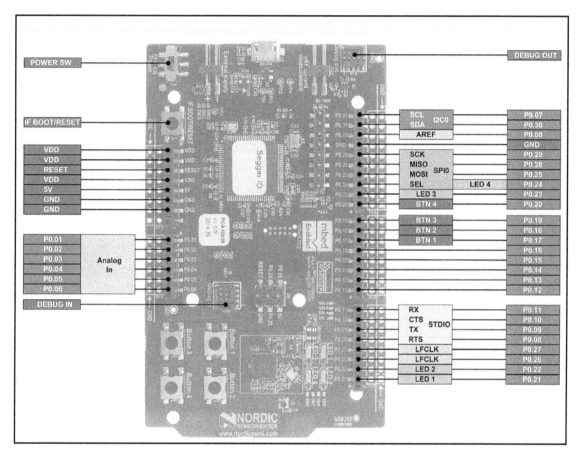

Nordic nRF51 DK Board with different components

Adafruit Bluefruit LE on nRF8001 Bluetooth Low Energy IC

Bluefruit LE is a Bluetooth Low Energy development kit that allows communication on a wireless link between Arduino and any BLE compatible device (iOS and Android included). Bluefruit works by simulating a **Universal Asynchronous Receiver/Transmitter (UART)** device under. The data transfer between the device and the wireless link is controllable by the developer and he can restrict or forward data as per the requirements. This device is equipped with nRF8001 chip by Nordic. nRF8001 is a BLE Connectivity Integrated Circuit with BLE v4.0 Radio, Link Layer and Host Stack. It also features a simple serial interface for external application microcontrollers. Adafruit takes advantage of this functionality and use nRF8001 as its brains.

Adafruit Bluefruit branding can be found here:

Nordic MicroBlue chip (nRF8001) logo

nRF8001 is designed for the applications that are for peripheral roles. That means it can be used in the applications like watches, remote controls and healthcare wearables and so on. This chip support GAP peripheral role out of the box. Some technical specs of this chip includes:

- 1.9-3.6V supply range
- DC to DC voltage regulator with 32-pins QFN package
- 2.4 GHz radio which is BLE compliant
- 1Mbps on-air data rate with -87 dBm receiver sensitivity

Adafruit clearly mentions in their documentation that this product is for advanced users who are comfortable using Nordic UART app, Bluefruit LE Connect app or can write his own iOS/Android application. The Bluetooth development kit looks something like this:

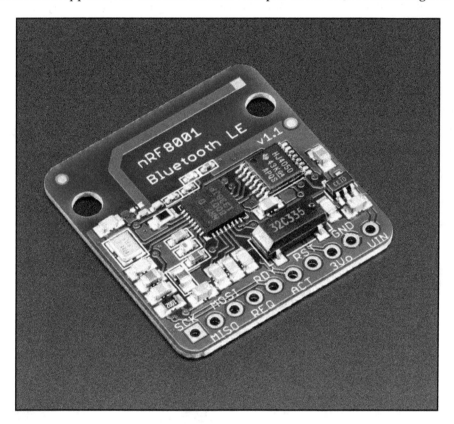

Adafruit Bluefruit LE Board with different components

 More information about nRF8001 Product Specification can be found on Nordic website
https://www.nordicsemi.com/eng/content/download/2981/38488/file/
nRF8001_PS_v1.3.pdf.

Coin Arduino BLE hardware

Coin BLE is an Arduino based hardware for iOS devices. The device is called coin because of its small dimensions. It measures only 0.7 x 2 inches in size and extremely low energy. The technical specifications of this hardware are:

- MCU: Atmel ATMEGA328AU with Arduino Pro Mini Bootloader
- Clock: 8MHz
- 32K flash and 2K RAM
- 3 to 18V power supply range, regulated to 3V
- Texas Instruments CC2540 F128 32MHz BLE chip with 128K flash and 8K RAM
- Johanson 2450AT18A100E Antenna

A comparison of Coin Arduino BLE kit bare board as compare to Assembled board can be found here:

From left to right: Bare Board, Assembled Board without Pins and Assembled Board with Pins

 More information about the use-cases of Coin Arduino can be found at this official video
`https://www.youtube.com/watch?v=qmugZcobahA&feature=player_embed ded`.

Bluetooth Low Energy software

Just like hardware, the development of Bluetooth Low Energy heavily relies on the software stack. Many software tools are available online for the development, testing and debugging of the BLE devices. Since Bluetooth Low Energy is heavily in correlation of the cellular phones, almost every major platform has its share in Bluetooth Low Energy software stack. PC industry is also very crucial in Bluetooth Low Energy software stack. In this section, we will emphasize more on the software/applications available on Microsoft Windows, Android and iOS.

Bluetooth Developer Studio by Special Interest Group

At this point of the book, we will talk about the software tools available for the development of Bluetooth Low Energy Applications. **Special Interest Group** (**SIG**) is the group responsible for maintaining and developing the new versions of Bluetooth. Their work also defines the future of the technology. That is why when SIG decided to reinvent the Bluetooth and move to the Bluetooth Low Energy, it was well-received by the industry.

Bluetooth Developer Studio is a software tool made by SIG that helps developer to learn Bluetooth technology. SIG claims that it reduces the education time by 50% and reduces time to market by 70%. This studio is helpful for GATT-based application development as developers can use it to generate Bluetooth profiles. It also provides various built-in profiles that describe many functionalities already built for BLE. If a developer wants a functionality already developed, he can grab the profile and implement in his application.

If a developer wants to implement a functionality that does not exist already, he can build hand-coded custom profiles using Bluetooth Developer Studio. The making and maintenance of these profiles are GUI based so if developer wants to implement a pre-written profile, he just need to drag and drop it. Once the view of these profiles are made and the description, GUID and other parameters are defined, developer studio provides many plugins to make it easy to add functionality which works on any device/chipset. The plugins also enable developers to export the code for the chips from Texas Instruments, Broadcom, Qualcomm, Arduino, Raspberry Pi and other developers.

This functionality simplifies the server side coding for the developer. Bluetooth Developers Studio also provides plugins that let you import the functionality to any Android/iOS app (as a client side application). Bluetooth Developer Studio simplifies the process of Bluetooth development and make it easily ported to any platform, firmware or software. This software is not that old, therefore many improvements are yet to made. Bluetooth Developer Studio also provides option to test the code using virtual or physical clients. The studio is supported by the Bluetooth developer's community that continuously adds new functionality (profiles/services) in the market.

Bluetooth Developer Studio comes with community supported plugins that give various additional functionalities. These are small JavaScript files that generates the code necessary to port your project on the platform of choice. We will discuss some Bluetooth Developer Studio plugins, to give reader an idea:

- **Frontline**: It allows developer to export the custom profiles in DecoderScript files which can be used in ComProbe Software.
- **Nordic nRF5x**: This plugin allows Bluetooth Development Studio to generate code for nRF52 and nRF51 series devices (nRF51 was discussed earlier in this chapter).
- **Matchbox**: It helps in Bluetooth client creation for any peripheral device. It also allows to make custom profiles without any code changes.
- **Renesas**: It helps to generate Bluetooth Profile/Service codes for Bluetooth Low Energy Protocol to run on Renesas microcontroller.

Bluetooth Developer Studio is available for free on Bluetooth SIG official website. `https://www.bluetooth.com/download-developer-studio`. On the other hand, the plugins can be downloaded directly from the software or from the website. A visual representation of the tool can be found here:

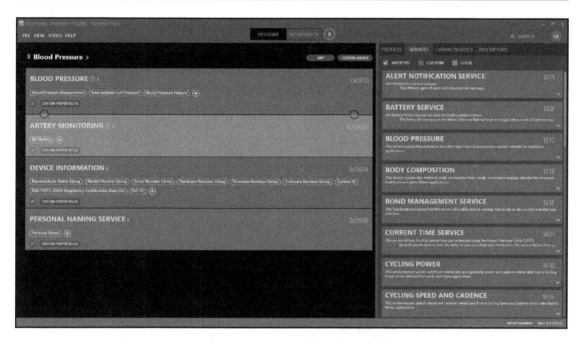

Bluetooth Developer Studio Interface with built-in services on the right hand side and Blood Pressure service on the left hand side with some characteristics.

Nordic Semiconductor tools

Nordic Semiconductor is a semiconductor company specializes in wireless system-on-chip solutions. Most of their wireless kits works on 2.4 GHz ISM band with low power consumption. They also specialize in Bluetooth Low Energy hardware especially nRF51 development kit that we discussed in this chapter. Their kit is available for developers to implement their own projects. In the software side, Nordic Semiconductor have many tools available for desktop and mobile platforms. These tools help developers to study, test and debug the Bluetooth connection and data exchange between central and peripheral. nRF tools also gives the functionality to perform over-the-air device firmware update.

Nordic targets popular platforms like Microsoft Windows, Android and iOS for their applications. This section of the book will specially focus on the Nordic's software suite and discuss each application and its functionality in detail.

nRF Connect for Mobile (Android and iOS)

nRF Connect formerly known as **nRF Master Control Panel**, is a powerful tool to scan, communicate and explore the nearby Bluetooth Low Energy devices. It also supports many Bluetooth SIG profiles including OTA DFU. There should not be any hesitation in declaring this application as one of the most powerful tools in the development of Bluetooth Low Energy applications. This is a free application available for Android and iOS devices. The latest branding is now:

nRF Connect logo by Nordic Semiconductors

nRF Connect provides the following powerful features for BLE:

- Scan and shows RSSI graph for each BLE device present in the proximity
- Parses advertisement and manufacturer specific data
- Connects to the BLE client and showing BLE services
- Parses characteristics for a BLE client
- Shows device information and BLE parameters
- Allows reading and writting characteristic
- Enables and receives BLE notifications and indications
- Allows creating GATT server with built-in configurations
- Over-the-air device firmware update
- Allows automated testing defined in XML file on BLE device
- Supports UART services
- Available for Android 4.3+ and iOS 8.0+

Following are the user-friendly screenshots of the nRF Connect app:

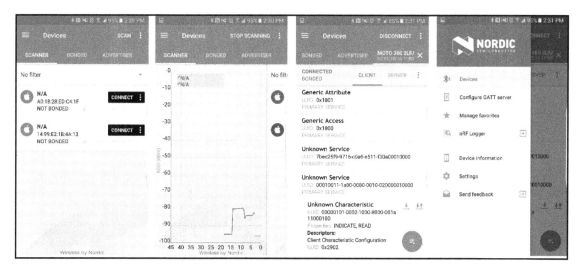

nRF Connect for Android with two devices, RSSI chart, services and app options

nRF Connect v4.10.0 is available to download on Play Store at `https://pl ay.google.com/store/apps/details?id=no.nordicsemi.android.mcp` and app v1.7 is available to download on Apple iTunes Store at `https ://itunes.apple.com/ca/app/nrf-connect/id1054362403?mt=8`.

nRF UART (Android and iOS)

Nordic Semiconductor's **Universal Asynchronous Receiver/Transmitter (UART)** app is used to connect BLE devices running a custom Nordic UART service. The app can send and receive ASCII/UTF-8 encoded data to the BLE device. UART in BLE context can be used to receive console outputs from the device. The app connects to the project `ble_uart_project_template` which is available at the GitHub repository of Nordic Semiconductor (`https://github.com/NordicSemiconductor/ble-sdk-arduino/tree/mas ter/libraries/BLE/examples/ble_uart_project_template`).

Arduino Bluetooth SDK works with BLE shielf for Arduino to setup the BLE link:

nRF UART icon for Android application

The app works with nRF2740 and nRF2741 modules. According to Nordic's documentation, the BLE shield is available from Seeed Studio and maker Shed UUIDs:

- **Service**: 6E400001-B5A3-F393-E0A9-E50E24DCCA9E
- **TX characteristics**: 6E400003-B5A3-F393-E0A9-E50E24DCCA9E
- **RX characteristics**: 6E400002-B5A3-F393-E0A9-E50E24DCCA9E

The UART app is available for Android 4.3+ and iOS 5.1+. The user experience of the UART app is very simple as described here:

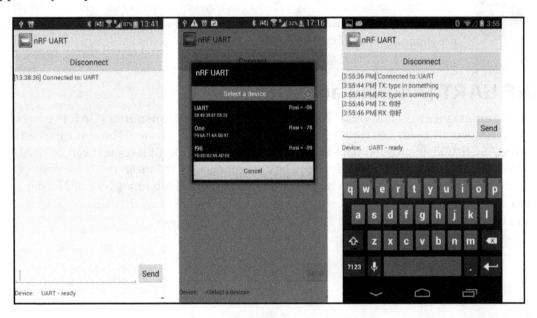

nRF UART app for Android with log screens and device selection screen

nRF UART app v2.0 is available to download on Play Store at `https://pl`
`ay.google.com/store/apps/details?id=com.nordicsemi.nrfUARTv`
`2` and app v1.0.1 is available at Apple's iTunes Store at `https://itunes.a`
`pple.com/ca/app/nrf-uart/id614594903?mt=8`.

nRF Logger (Android)

nRF Logger is a good utility application for Bluetooth Low Energy communication. It
allows users to view the log sessions during the communication process. This utility is used
by other Nordic applications as well. Logging the sessions bring flexibility in debugging.
Developers can debug their applications using six log levels: debug, verbose, info,
application, warning and error. These debug level are very similar to the Android
debugger, where you can log the events in one of these categories. In the next screenshot,
you can see the simple interface of the logger and how it can be useful for debugging
purposes:

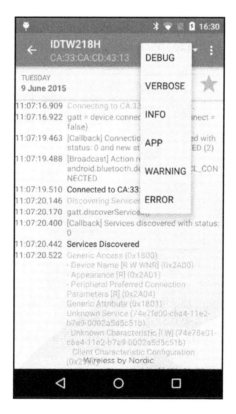

nRF Logger application interface

This application is now an open source Android project on GitHub in the form of a library. Developers can create log entries from their application directly into the nRF Logger applications. Nordic tried to make this logger like Android debugger to give a similar environment to the developers. nRF logger is available for Android 4.1+. Nordic made it extremely easy to integrate logger library in any Android application:

1. Include compile `no.nordicsemi.android:log:2.0.0` in your `gradle.build` file.

2. Make a new logging session by writing `mLogSession = Logger.newSession(getActivity(), key, name);` in your activity file.

3. Log the values and errors using `Logger.log(mLogSession, Level.INFO, text);` and `Logger.e(mLogSession, R.string.error, someArg);` in your activity file.

4. nRF Logger session description can also be set in the Android code by using `Logger.setSessionDescription(mLogSession, "SET DESCRIPTION HERE");` in your activity file.

5. `Logger.setSessionMark(mLogSession, Logger.MARK_FLAG_RED);` can be used in your activity file to mark the session with one of six symbols.

nRF Logger application icon on Android

The key variable is used to group the logs of one day together. Another important thing to remember is that if you want to use nRF Logger in your application, you need to pre-install it. If the application is not pre-install in the cellular phone, it will ask to download the application first.

The GitHub repository of nRF Logger can be found at
`https://github.com/NordicSemiconductor/nRF-Logger-API`.

nRF Beacon for Eddystone (Android and iOS)

This application is developed to support GATT configuration services for Eddystone beacons. This allows user to configure the beacon to advertise all types of Eddystone frame types. These include UID, URL, TLM, EID and eTLM frame types. Eddystone-UID frame broadcast a unique 16-byte beacon ID which is a collection of 10-byte namespace and 6-byte instance.

Eddystone-EID on the other hand is a component of Eddystone protocol which broadcast an identifier every minute. The identifier is resolved in useful information by a special service that shares a key (**ephemeral identity key** or **EIK**) with individual beacon.

The Nearby scan is followed by UID/EID registration and create/retrieve attachments for them on the Proximity API cloud. Proximity Beacon API is a cloud service that let you manage data associated with your beacon using REST interface. The branding of this app can be found in the following figure:

nRF Beacon for Eddystone icon by Nordic Semiconductors

Moving forward from the basic introduction of beacons, the nRF beacon for Eddystone application consists of following features:

- Beacon scanning using Nearby API with registration on Proximity cloud service
- EID registration on the Proximity cloud and data retrieval

- Configuring Eddystone beacons using GATT configuration service
- Google URL shortener to configure URL beacons

A view of the nRF beacon app can be found below. It shows a beacon in proximity with an attachment of **Welcome to Nordic Semiconductor**:

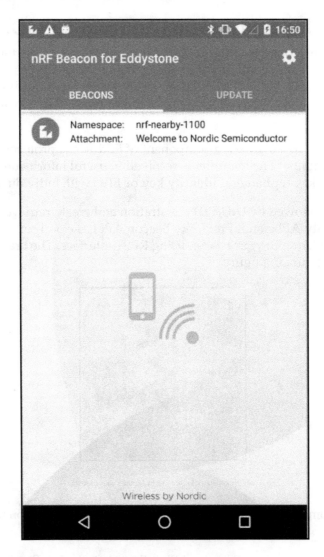

nRF Beacon for Eddystone main interface

The source code of nRF Beacon for Eddystone is available on GitHub as an open source project. It requires Android 4.3+ with Access Coarse Location permission.

 nRF open source GitHub repositories can be found on `https://github.com/NordicSemiconductor/nrf5-sdk-for-eddystone,` and `https://github.com/NordicSemiconductor/Android-nRF-Beacon-for-Eddystone.`

nRF BLE Joiner (Android)

This is relatively a new application from Nordic Semiconductors. It is used to add new IoT nodes to a IPv6 enabled router via Bluetooth Smart. Firstly, it scans for IoT nodes in range which contains BLE Node Configuration Service (a Nordic propriety service). Secondly, it will connect to the node and add these nodes to the network based on Bluetooth Smart. You can add your Wi-Fi network in the application to connect the BLE nodes. You can manually install or scan for an existing Wi-Fi network.

 The application is open-source and the code can be found at `https://github.com/NordicSemiconductor/Android-nRF-BLE-Joiner.` The application needs Android 4.3+ with Access Coarse Location permission.

The configuration of a node in the nRG BLE Joiner app can be seen here with some parameters:

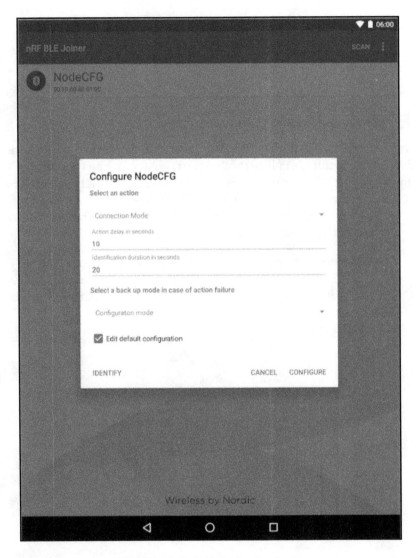

nRF BLE Joiner Application interface for node configuration

Google Beacon tools (Android)

Beacon Tools is an application developed by Google Inc. and is used to register Bluetooth low Energy beacons. It uses Google Registry and create attachments for them. This application can be used to scan and register the Bluetooth beacons. So when the beacons are scanned from a beacon application, it can serve them with registered attachments.

The application gives an attraction interface with the position of the beacon pointing at the Google map with parameters. It shows all the nearby beacons and maintains a list of their UIDs. The Google beacon tool gives a comprehensive view of all the registered beacons with UIDs in the hex form. The application was launched in April and still in the early stage of its development.

Figure describes the user interface of the application with Google Maps pointing at the location of the beacon:

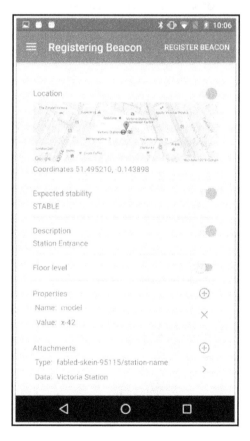

Google Beacon Tools Application with registering beacon interface

Physical web (Android)

This is an open-source application developed by the physical web team in Google. The concept of physical web is introduced by Google where it is proposed that instead of writing the URL in your browser, the IoT device should tell the URL and allow a user to open a web-link. This is done by a discovery service powered by Bluetooth Low Energy Beacons.

A use-case for the physical web can be understood by considering a smart parking meter. Imagine if the parking meter is continuously broadcasting a URL to control/pay the parking ticket, the driver just need to take out the cellphone, accept the URL and pay directly without any physical interaction or worry of downloading any app. The physical web app also lets you manage your own beacon and let you edit the broadcasting URLs.

In the figure below, you can see that the application is searching for the nearby beacons. If the beacons are available, their broadcasted value can be seen in the applications:

Google physical web Android searching for the beacons

The application for physical web is available in the Play store `https://play.google.com/store/apps/details?id=physical_web.org.p hysicalweb`. The application is an open-source project. Thus, the code can be found at the Google GitHub repository `https://github.com/google/p hysical-web/`.

Summary

The focus of this chapter was to give readers an idea about the hardware and software tools available for the development. It described some examples of the Bluetooth hardware available for development including Nordic nRF51 development kit (with soft-devices), Adafruit Bluefruit BLE development kit and Coin Arduino BLE kit. In the software category, the chapter discussed Bluetooth Developer Studio by SIG, Nordic Semiconductor tools like nRF Connect, nRF Logger, nRF Beacon for Eddystone, nRF BLE Joiner and nRF UART 2.0. The chapter concluded itself with the introduction of two beacon applications by Google. Google Beacon Tools and Google physical web which are used in the Internet of Things applications. The chapter emphasized on the technical specifications and the uses of these tools rather than the explanation of the functionality. You will learn how to use these tools to their best in the coming chapters.

In the next chapter we will discuss the core of Bluetooth Low Energy communication, the Central and Peripheral Communication model with examples.

3

Building a BLE Central and Peripheral Communication System

Bluetooth Low Energy is a comparatively recent communication technology that is spreading wildly in our world. Due to low power consumption, it is easily implementable on cellular platforms, thus providing millions of users a fast and secure way to transfer their data from one device to another. In the previous chapters, we discussed the origin of Bluetooth Low Energy, its usage in the Internet of Things and some software/hardware development kits that can be used to implement your personal Bluetooth Low Energy device. From now on, we will start to put things in practice and implement things with the tools discussed in the last chapter. This chapter will define the very core of the Bluetooth Low Energy communication model by discussing:

- An overview of BLE Central/Peripheral architecture with the role of GATT Server/Client
- Android platform for Bluetooth Low Energy Communications
- Android services and characteristics using UUIDs
- Bluetooth pairing and bonding
- Android application to list Bluetooth devices in close proximity
- Heart rate monitor application on Android
- Over-the-air **Device Firmware Update** (**DFU**) on Android

Bluetooth Low Energy central and peripheral

The fundamental communication technique used in **Bluetooth Low Energy** (BLE) is client-server architecture. This mode of communication has been used in the computing industry for over half a century and is proven to be very successful over the years. The term client-server is used in its critical definition:

- **Client**: A device that asks for the data from the server
- **Server**: A device that serves the data to the client

It is a distributed architecture that divides the responsibilities between a receiver and a provider, in other words, a client and a server. Technically, client and server reside on different machines but it is completely possible that single-machine hosts both. Bluetooth Low Energy works on the same concept of data exchange. There is a BLE GATT client which asks for the data from the BLE GATT server.

 The word GATT is an acronym of the Generic Attribute Profile and it defines the way a BLE device behave in the communication.

Keeping GATT in mind, we will now discuss the concept of Peripheral-Central communication in BLE. Peripheral is a device that advertises itself and let other devices (centrals) know about their existence. Central, upon hearing the notification initiates the connection request. These terms are defined solely on the basis of the connection request and advertisement and they are not related to the client and server roles of the BLE communication. Although there is no connection in these terminologies, peripherals mostly serve as servers and centrals as clients. These roles are easily exchangeable based on the type of application.

 Sometimes, a central is referred as a master and a peripheral as a slave.

A non-connecting variant of Peripheral and Central are called **Broadcaster** and **Observer**, respectively. Just like central and peripheral, these terminologies are also defined by Bluetooth Low Energy Core Specifications and often used in the literature. Broadcaster advertises itself by sending advertisement messages while observer hears the advertisement and receives the scanned data. These are non-connecting roles that means that the broadcaster and observer will never connect to each other for further data exchange.

In order to put things in black and white, refer to the following diagram:

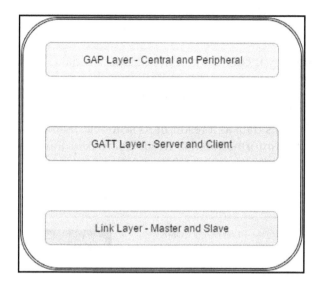

Terminologies defined based on the layers

Examples of Bluetooth central peripheral communication

This section will consider some examples of central-peripheral communication in Bluetooth Low Energy so that the readers get a clear vision of the possibilities. As described before, central and peripherals are the terms used in accordance with the GATT roles. The advertiser is peripheral and the connection initiator is central. Peripheral will broadcast its identity but once it is paired with a central, its identity will only be recognized by that central until unpaired. This function gives better security to BLE peripherals because they will not be shown in the open scan by other devices.

In central-peripheral communication, centrals do not advertise their existence (unless it is a feature in the cellular-phone). They can scan and read the advertisements from the peripherals before connecting to it. Advertisement data can be configured on peripheral so that the peripheral can broadcast useful information to the central without any need of a connection. A sample advertisement data from peripheral includes:

- Advertisement type
- Transmission power

- Service UUID and relevant data
- Device's physical address
- Flags (such as GeneralDiscoverable and LeAndBrErdCapable and so on)

The advertisement data may include manufacturer specific data. There is no limitation on the size of this data but it is recommended to limit it. If the advertisement data is long, it may compromise the power efficiency of the device broadcasting it. The manufacturer's data contains its id which is typically 2 bytes long which is followed by the manufacturer's data. Central can receive this data as a result of its scan and then parse it to extract information. Any sensitive information that needs not to be publicly available should not be put in the manufacturer's data.

A view of the manufacturer's data in Nordic Connect app is shown as follows:

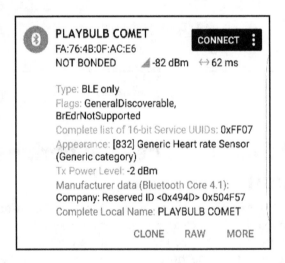

A view of manufacturer's data in Nordic Connect app

 Readers can download nRF Connect app and try scanning the surrounding BLE devices to study their advertising data.

Smartphone – smartwatch communication

Probably the most common use-case of BLE Central-Peripheral is the communication model between a smartphone and a smartwatch. Smartwatches are becoming very popular in our world.

Their primary purpose is to give controls over various apps (such as WhatsApp, Messenger, and Facebook etc.) without taking out the smartphone. In most cases, smartwatch comes as an accessory for a smartphone as it is dependent on the phone to receive and send data. Some manufacturers are making standalone smartwatches, which brings their own sensors (such as Wi-Fi and GPS etc.) to reduce the dependencies from the smartphone.

In order to understand the initial communication between a smartphone and a smartwatch, we will consider a real life example of a Motorola 360 smartwatch and a Samsung Galaxy S7 Edge smartphone. When a smartwatch comes out of the box, it needs a connection from a smartphone. It broadcasts its identity to everyone in the proximity performing the typical peripheral behavior. At this point, any smartphone (or central) can initiate the connection request to the smartwatch.

Moto 360 first asks to download and install Android Wear on the smartphone in order to fulfil the connection process. Once the software is installed, the pairing process is started. Smartwatch broadcasts its identity as Moto 360 3LFJ, which is visible on the smartphone as well. Upon pressing it, smartphone and smartwatch both show a 7-digit passkey (such as 7066595) which needs to be confirmed by the user. This step helps the user to verify that the user is pairing correct smartwatch with the smartphone. The devices are now paired with each other.

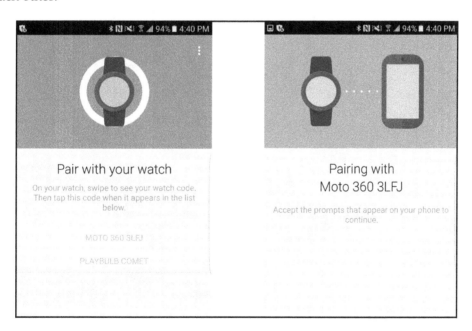

Left: Peripheral is broadcasting its identity Moto 360 3LFJ and Right: Central-Peripheral pairing is in process

This is the typical process of pairing a central to its peripheral. When both ends are properly paired with each other, the peripheral does not broadcast its identity publicly. Even if the peripheral is disconnected from the central, peripheral will only search for that particular central while ignoring all other centrals in the proximity.

Smartphone – smart LED strip communication

Another example of central-peripheral communication can be seen in the form of a smartphone to smart LED strip. The process is the same as the previous example. LED strip (peripheral) broadcasts its identity publicly. This broadcast can be heard by any scan call in the proximity. Smartphone (central) needs to initiate the connection request before sending any command. Once they are paired, central can send commands to the peripheral and control the color, brightness and give effects to the light. All these commands are initiated by the central and received by the peripheral.

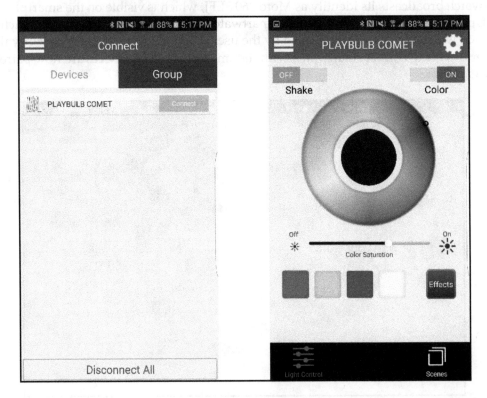

Left: PlayBulb LED Strip (peripheral) is broadcasting its identity PLAYBULB COMET and Right: Connection is successful and central can send a command to the peripheral

Android Bluetooth Low Energy

This book promises a theoretical as well as a practical approach towards the Internet of Things. From this point onwards, readers will understand the practical aspects of the technology. This section deals with the Android API for Bluetooth Low Energy. Although we are taking Android as a medium to study Bluetooth Low Energy for the most part of the book, we do realize the important of iOS platform. Hence, we will also discuss some SWIFT code in the upcoming chapters.

 The reason why Android is given an edge over iOS is because it is widely spread. According to the study by Patently Apple, 81% of the smartphones in the world runs Android. We wanted to make sure that many people may take advantage from this book.

An introduction to Android

Android is an open-source mobile platform based on Linux kernel and developed by Google Inc. It was primarily developed for smartphones but now exists in many variants including Android Wear, Android Auto, and Android TV. The open-source nature of this operating system leads it to be adopted by many multi-national companies like Samsung, HTC, LG and Sony etcetera for their smartphones. These smartphone manufacturers sometime write their own user interface over Android and call it a skin. For example, Samsung calls their UI "TouchWiz" and HTC call theirs "HTC Sense".

The initial release of Android came out eight years ago in September 2008. Since then, it evolved significantly through the versions. Interestingly, Android version names are all in alphabetical order and all named after some kind of sweet. It is not abnormal to say that developers do have their sweet-tooth, which is proved by Google. A complete list of Android Operating System is given here:

- Android Nougat v7.0 released on August 22, 2016
- Android Marshmallow v6.0 released on October 5, 2015
- Android Lollipop v5.x first released on November 3, 2014
- Android KitKat v4.4 released on October 31, 2013
- Android Jelly Bean v4.x first released on July 9, 2012
- Android Ice Cream Sandwich v4.0 released on December 16, 2011
- Android Honeycomb v3.x, first released on February 22, 2011
- Android Gingerbread v2.3 released on February 9, 2011
- Android Froyo v2.2 released on May 20, 2010

- Android Éclair v2.0 - v2.1, first released on January 12, 2010
- Android Donut v1.6 released on September 15, 2009
- Android Cupcake v1.5 released on April 30, 2009
- Android Banana Bread v1.1
- Android Apple Pie v1.0

The current version of Android is called Android Nougat (v7.0) and will be used as a benchmark for this book.

Android Nougat v7.0 official logo

Bluetooth Low Energy in API level 24

In this section, we will discuss some important Android BLE classes and interfaces in API level 24 (made for Android Nougat 7.0). These classes and interfaces are essential for BLE development in Android as they give a way to access Bluetooth LE hardware efficiently. The main tasks like scanning, connecting, implementing the GATT client and GATT server roles and transferring data are achieved through these classes. We assume that the reader knows the basics of Android development. Nevertheless, we will still explain things in the fairly understandable way to allow beginners to catch up the pace.

Bluetooth Low Energy permissions

In Android project structure, there is a file named `AndroidManifest.xml` which contains a high level overview of the Android application. This file contains the information about the activities, services, receivers, providers, libraries, and permissions. It is essential to give explicit permissions of Bluetooth. This can be achieved by providing the following line of code in the Manifest file:

```
<uses-permission android:name="android.permission.BLUETOOTH" />
<uses-permission android:name="android.permission.BLUETOOTH_ADMIN" />
```

This permission ensures that the application is able to access Bluetooth. Another important permission for the proper execution of Bluetooth Low Energy scanning is:

```
<uses-permission android:name="android.permission.ACCESS_COARSE_LOCATION"
/>
<uses-permission android:name="android.permission.ACCESS_FINE_LOCATION" />
```

Starting Android Marshmallow v6.0, it is essential to give coarse and fine location permissions to start BLE scan. It is because Bluetooth Low Energy scanning works in the background and is blocked unless these permissions are given. This permission needs to be given in the Manifest file as well as on runtime. To request the location permission at runtime you need to write:

```
// ---------------------- MANUAL PERMISSIONS --------------------------
String[] permissionString = new
String[]{Manifest.permission.ACCESS_COARSE_LOCATION,
Manifest.permission.ACCESS_FINE_LOCATION};
yourActivity.requestPermissions(permissionString,
yourPermissionRequestCode);
// ---------------------- CATCHING PERMISSIONS ------------------------
@Override
public void onRequestPermissionsResult(int requestCode, String
permissions[], int[] grantResults){
    if(requestCode == yourPermissionRequestCode)
    {
        ... //Do something based on the user click.
    }
}
```

This will ensure that the user is notified to give coarse and fine location permission before the BLE process starts. Since this permission is only required for Android Marshmallow and higher version, it is a good practice to put the previously mentioned code in a version check.

Bluetooth Low Energy interfaces

There are three kinds of interfaces given for BLE in Android API 24. We will now introduce those interfaces in the following sections.

BluetoothAdapter.LeScanCallback

An interface required delivering the results of LE Scan. You need to implement the method `onLeScan()` which will return back scanned device, its RSSI (signal strength) and the scan record array. Later this object will be passed to the `startLeScan` method.

```
BluetoothAdapter.LeScanCallback leScanCallback = new
BluetoothAdapter.LeScanCallback() {        @Override        public void
onLeScan(BluetoothDevice device, int rssi, byte[] scanRecord) {           //
Do something with the scanned data...       } };
startLeScan(leScanCallback);
```

BluetoothProfile.ServiceListener

This interface is used to notify the `BluetoothProfile` clients when they are connected or disconnected from the service. The must implemented methods for this interface are `onServiceConnected()` and `onServiceDisconnected()`.

```
BluetoothProfile.ServiceListener leServiceListener = new
BluetoothProfile.ServiceListener() {      @Override       public void
onServiceConnected(int profile, BluetoothProfile proxy) {          //Do
something upon receiving a service connection      }     @Override
public void onServiceDisconnected(int profile) {         //Do something
upon receiving a service disconnection       } };
```

Bluetooth Low Energy classes

We will now discuss some important classes for Android BLE. These will help readers to understand the upcoming examples.

BluetoothAdapter

This is the fundamental class which lets you accomplish basic Bluetooth tasks such as device discovery etc. You can get the adapter by calling:

```
// Initializes Bluetooth adapter. final BluetoothManager bluetoothManager =
(BluetoothManager) getSystemService(Context.BLUETOOTH_SERVICE);
mBluetoothAdapter = bluetoothManager.getAdapter();
```

This `BluetoothAdapter` is universal and the application can interact with it to transfer and receive data over the Bluetooth link. You can also find methods like `startLeScan()` and `stopLeScan()` in this class, which will then be implemented using classes of `LeScanCallback`interface.

BluetoothGatt

This class was added in Android API Level 18 and it enables the basic Bluetooth Smart communication. Functionalities like service discovery and reading/writing on characteristics are achieved using this class. In order to use this class, you first need to get the `BluetoothGatt` object by calling `connectGatt()` method on the scanned device.

```
BluetoothGatt mBluetoothGatt = device.connectGatt(this, true,
mBluetoothGattCallback);
```

This `mBluetoothGatt` object can now be used to start service discovery process and read/write the characteristics. In order to connect to the device, you need to describe a BluetoothGattCallback class which will ask you to override methods to do tasks when a certain event triggers. We will discuss this class in the next section but for now, a short view of the methods offered by BluetoothGatt class can be seen here:

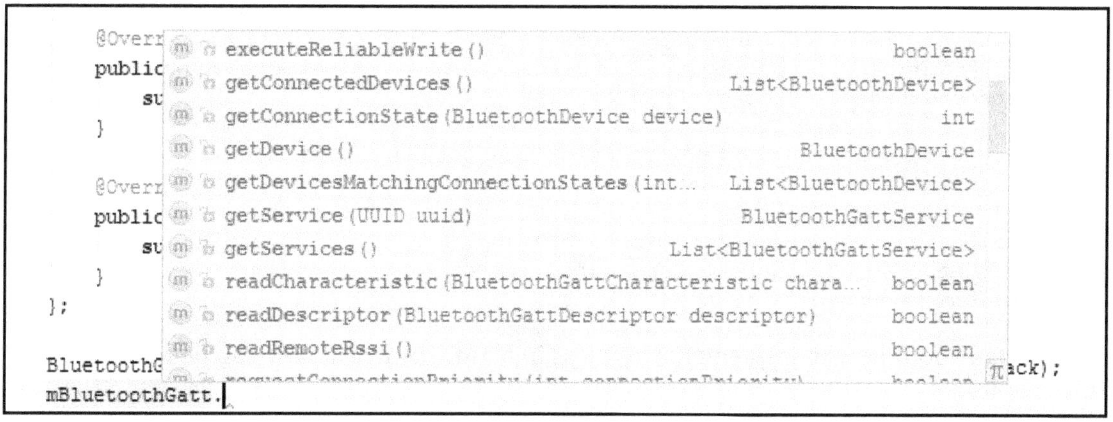

IntelliSense offered by Android Studio to list BluetoothGatt methods

BluetoothGattCallback

This is an important class in the Bluetooth communication which gives an interface to the developer to describe custom tasks performed upon a triggered event. For example, by implementing an onServicesDiscovered method, we can define the tasks to be performed after all the services are discovered. Similarly, onCharacteristicRead can be overridden to perform tasks upon reading a characteristic.

```
BluetoothGattCallback mBluetoothGattCallback = new BluetoothGattCallback()
{       @Override     public void onConnectionStateChange(BluetoothGatt gatt,
int status, int newState) {              super.onConnectionStateChange(gatt,
status, newState);       }      @Override     public void
onServicesDiscovered(BluetoothGatt gatt, int status) {
super.onServicesDiscovered(gatt, status);       }      @Override     public
void onCharacteristicRead(BluetoothGatt gatt, BluetoothGattCharacteristic
characteristic, int status) {              super.onCharacteristicRead(gatt,
characteristic, status);       }      @Override     public void
onCharacteristicWrite(BluetoothGatt gatt, BluetoothGattCharacteristic
characteristic, int status) {              super.onCharacteristicWrite(gatt,
characteristic, status);       }      @Override     public void
onCharacteristicChanged(BluetoothGatt gatt, BluetoothGattCharacteristic
characteristic) {            super.onCharacteristicChanged(gatt,
characteristic);       }      @Override     public void
onDescriptorRead(BluetoothGatt gatt, BluetoothGattDescriptor descriptor,
int status) {            super.onDescriptorRead(gatt, descriptor, status);
}      @Override     public void onDescriptorWrite(BluetoothGatt gatt,
BluetoothGattDescriptor descriptor, int status) {
super.onDescriptorWrite(gatt, descriptor, status);       }      @Override
public void onReliableWriteCompleted(BluetoothGatt gatt, int status) {
super.onReliableWriteCompleted(gatt, status);       }      @Override
public void onReadRemoteRssi(BluetoothGatt gatt, int rssi, int status) {
super.onReadRemoteRssi(gatt, rssi, status);       }      @Override     public
void onMtuChanged(BluetoothGatt gatt, int mtu, int status) {
super.onMtuChanged(gatt, mtu, status);       } };
```

More detail about how to implement this class in a more practical way can be found later in this chapter.

BluetoothGattService

This class is responsible for holding different parameters and functionalities related to the Bluetooth Service. `BluetoothGattService` gives flexibility to the developer to get the characteristics related to that service which is significant for writing and reading data to and from the Bluetooth characteristics. A quick view of the methods offered by this class can be found here:

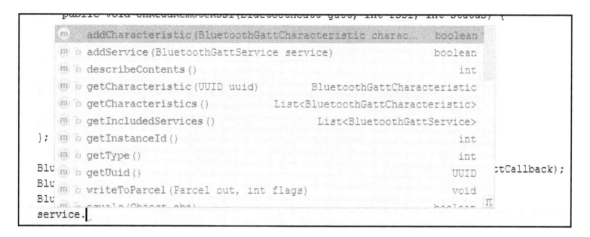

IntelliSense offered by Android Studio to list BluetoothGattService class.

A service can be acquired by calling `getService` method on `mBluetoothGatt` object after connecting to a particular device:

```
BluetoothGattService service =
mBluetoothGatt.getService(UUID.randomUUID());
```

An example of getting a random characteristic from a service object is:

```
BluetoothGattCharacteristic characteristic =
service.getCharacteristic(UUID.randomUUID());
```

 Note that the `randomUUID()` method is used for demonstration purpose and in the real-world application this UUID has to match the peripheral's UUID.

BluetoothGattCharacteristic

This class contains the functionality and parameters of a Bluetooth Characteristics. Methods related to reading and writing the data can be found in this class. Normally, a list of characteristics is acquired from the service object.

```
List<BluetoothGattCharacteristic> gattCharacteristics =
service.getCharacteristics(); // Loops through available Characteristics.
for (BluetoothGattCharacteristic gattCharacteristic :
gattCharacteristics) {      // Do something with gattCharacteristic }
```

And a view of the functionalities offered by this class can be found here:

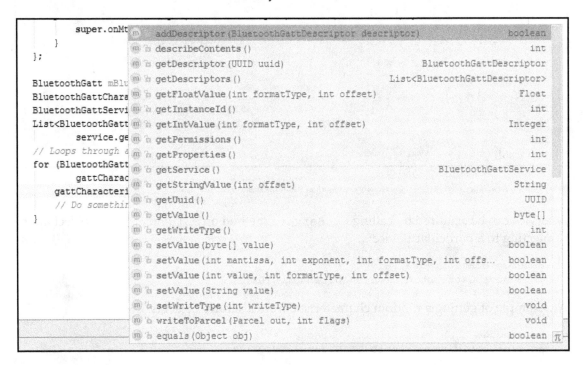

IntelliSense offered by Android Studio to list BluetoothGattCharacteristic methods

Building an Android app to list Bluetooth devices in the proximity

In this section, we will make an Android application that will display a list of Bluetooth devices in the proximity. The example will make sure that it covers each and every step necessary to build a bluetooth application, including managing the permissions and asking the permissions in the real time. The main purpose of this book is to give a hands-on experience to the user hence, we are assuming that the reader has basic Android knowledge and will be able to make a project to start things.

Conventions

The writing convention used in this book makes sure that it is simple for the reader and we highly encourage readers to adopt a more meaningful naming conventions for their projects. The code attached with this section takes `BluetoothScanExample` as project name and `com.demo.bluetoothscanexample`as the package name. `MainActivity.java` class is used as the main class which will list the bluetooth devices.

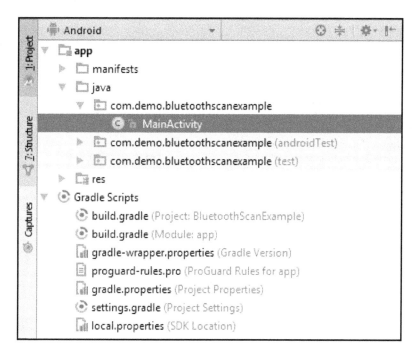

Android Studio views showing Naming Conventions for the project

Permissions in AndroidManifest.xml

We will first start off by giving Bluetooth permissions in the `AndroidManifest.xml` file. These permission includes:

```
<uses-permission android:name="android.permission.BLUETOOTH" /> <uses-
permission android:name="android.permission.BLUETOOTH_ADMIN" /> <uses-
permission android:name="android.permission.ACCESS_FINE_LOCATION" /> <uses-
permission android:name="android.permission.ACCESS_COARSE_LOCATION" />
```

Location permissions are necessary to run the scan, as the process requires to be run in the background and from Android Marshmallow onwards, any background process needs location permission otherwise, it will not yield the same result or no result.

Since we are developing the application for Bluetooth Low Energy, it is a good practice to include a `<uses-feature>` tag to mark Bluetooth LE as a required feature of the phone.

A view of the `AndroidManifest.xml` file is:

```
<?xml version="1.0" encoding="utf-8"?> <manifest
xmlns:android="http://schemas.android.com/apk/res/android"
package="com.demo.bluetoothscanexample">      <uses-permission
android:name="android.permission.BLUETOOTH" />      <uses-permission
android:name="android.permission.BLUETOOTH_ADMIN" />      <uses-permission
android:name="android.permission.ACCESS_FINE_LOCATION" />      <uses-
permission android:name="android.permission.ACCESS_COARSE_LOCATION" />
<uses-feature android:name="android.hardware.bluetooth_le"
android:required="true"/>      <application
android:allowBackup="true"          android:icon="@mipmap/ic_launcher"
android:label="@string/app_name"          android:supportsRtl="true"
android:theme="@style/AppTheme">          <activity
android:name=".MainActivity">          <intent-filter>
<action android:name="android.intent.action.MAIN" />
<category android:name="android.intent.category.LAUNCHER" />
</intent-filter>          </activity>      </application> </manifest>
```

Another good practice is to check if the device supports Bluetooth Low Energy at the `onCreate()` Method of the main activity. This can be achieved by putting this in the `onCreate()` method:

```
if
(!getPackageManager().hasSystemFeature(PackageManager.FEATURE_BLUETOOTH_LE)
) {      Toast.makeText(this, R.string.ble_not_supported,
Toast.LENGTH_SHORT).show();      finish();
}
```

Runtime permissions

In addition to the manifest permissions, Android needs explicit runtime permissions for Bluetooth and Coarse Location. Irrespective of the application flow, these permissions should be asked at the start of the application lifetime. To get the coarse location permission, you can add:

```
if
(this.checkSelfPermission(android.Manifest.permission.ACCESS_COARSE_LOCATIO
N) != PackageManager.PERMISSION_GRANTED) {      requestPermissions(new
String[]{android.Manifest.permission.ACCESS_COARSE_LOCATION},
REQUEST_CODE_COARSE_PERMISSION);
}
```

The method `checkSelfPermission` confirms if the location permission is not already given. On the other hand, the `REQUEST_CODE_COARSE_PERMISSION`is a static integer which can be caught by overriding `onRequestPermissionResult()` method:

```
@Override public void onRequestPermissionsResult(int requestCode, String[]
permissions, int[] grantResults) {
super.onRequestPermissionsResult(requestCode, permissions, grantResults);
    if(requestCode == REQUEST_CODE_COARSE_PERMISSION){
Toast.makeText(this, "Coarse Permission Granted...",
Toast.LENGTH_LONG).show();      }
}
```

The UI dialog for location permission will look like this:

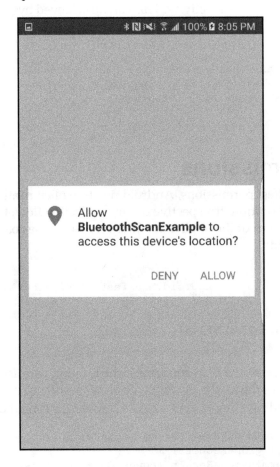

User Interface for location permission

Following the location permission, the next permission is the Bluetooth permission. It is the most important permission and for all the Android examples in this book. In order to give the permission, you need to add following code:

```
if (!mBluetoothAdapter.isEnabled()){
    Intent enableBtIntent = new
Intent(BluetoothAdapter.ACTION_REQUEST_ENABLE);
    startActivityForResult(enableBtIntent,
REQUEST_CODE_BLUETOOTH_PERMISSION);
}
```

mBluetoothAdapter is a global instance of the class BluetoothAdapter. isEnabled() method tells if the Bluetooth is enabled in the cellphone. If the Bluetooth is not enabled, we can call a pre-defined intent by using ACTION_REQUEST_ENABLE. All the startActivityForResult() calls are returned back to the activity and can be caught in onActivityResult();

```
@Override protected void onActivityResult(int requestCode, int resultCode,
Intent data) {       super.onActivityResult(requestCode, resultCode, data);
if(requestCode == REQUEST_CODE_BLUETOOTH_PERMISSION){            // Do
something       }
}
```

The Bluetooth permission dialog will look like this:

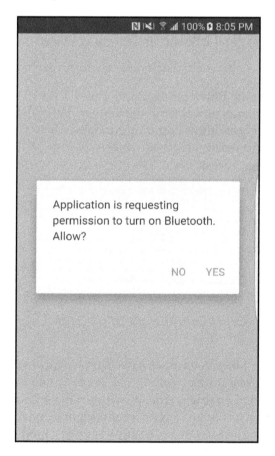

User Interface for Bluetooth permission

Once these runtime permissions are given, the next step is to start the scanning process.

Bluetooth scanning process

For the scanning process, following objects should be prepared in order:

```
BluetoothManager mBluetoothManager =           (BluetoothManager)
getSystemService(Context.BLUETOOTH_SERVICE); BluetoothAdapter
mBluetoothAdapter = mBluetoothManager.getAdapter();
```

`getSystemService()` method brings back the Bluetooth Service Manager, if called with `Context.BLUETOOTH_SERVICE`. This manager can then be used to get the adapter which is the core of the Bluetooth operation. As mentioned in the code, `mBluetoothAdapter` can be acquired by calling `getAdapter()` on the `BluetoothManager` object. This bluetooth adapter will now be used to get the BluetoothLeScanner object:

```
mBluetoothLeScanner = mBluetoothAdapter.getBluetoothLeScanner();
```

It is a good practice to start the Bluetooth scanning process for 10 to 15 seconds so that the phone don't consume continous power from the battery. This purpose will be achieved by Android Handler. Handler class allows sending runnable objects associated with the thread's `MessageQueue`. We wrote a method which will run for 10 seconds and will provide Bluetooth scanning process:

```
private void scanBluetoothDevices(boolean enable) {      if
(mBluetoothLeScanner != null) {              if (enable) {              // Stops
scanning after a pre-defined scan period.
mHandler.postDelayed(new Runnable() {                    @Override
public void run() {                        mScanning = false;
mBluetoothLeScanner.stopScan(mBluetoothScanCallBack);                    }
}, 10000);            mScanning = true;
mBluetoothLeScanner.startScan(mBluetoothScanCallBack);          } else {
mScanning = false;
mBluetoothLeScanner.stopScan(mBluetoothScanCallBack);              }      }
}
```

`scanBluetoothDevice()` takes boolean as a decision to start or stop the scan. If `enable` is false, the scanner will stop scanning by calling `mBluetoothLeScanner.stopScan()` method. On the other hand, if **enable** is true, it will start the scanning by calling `mBluetoothLeScanner.startScan()` with a callback object. Let's forget the callback for a bit time and concentrate on the `mHandler.postDelayed()` method.

`mHandler.postDelayed(Runnable, x)` is a method executes the provided runnable after x amount of time. In our case, we are providing 10000 millisec as the processing time. The runnable will be executed after 10 seconds which will interupt the scanning process and stop the scan by calling `mBluetoothLeScanner.stopScan()`.

`mBluetoothScanCallback` is an object of `ScanCallback` which provides developers methods for overriding. The most important methods of this class are: `onScanResult` and `onScanFailed`. When the scan yields any result the `onScanResult` is triggered and the developer can update the user interface. `onScanFailed`, on the other hand, is called if for some reason the scan fails. You can update the view in this case as well.

In our application, we are updating the list view adapter by giving the scanned Bluetooth device to the adapter and calling `notifyDataSetChanged()`.

```
private ScanCallback mBluetoothScanCallBack = new ScanCallback() {
@Override     public void onScanResult(int callbackType, final ScanResult
result) {          super.onScanResult(callbackType, result);
runOnUiThread(new Runnable() {              @Override          public
void run() {
mLeDeviceListAdapter.addDevice(result.getDevice());
mLeDeviceListAdapter.notifyDataSetChanged();              }         });
}     @Override     public void onScanFailed(int errorCode) {
super.onScanFailed(errorCode);          Toast.makeText(MainActivity.this,
"Scan Failed: Please try again...", Toast.LENGTH_LONG).show();      }
};
```

`result.getDevice()` will return a `BluetoothDevice` object which can then be passed to the custom list adapter for UI update. In previously mentioned code, we are adding the device to the list adapter and calling `notifyDataSetChanged()`. This method is used to notify the observer that the underlying data is being changed and the view should update itself.

Custom list adapter and ListView

For the UI purposes, it is good to have a custom list adapter that shows the name and address of the scanned Bluetooth device in the list. For this purpose, we created a custom adapter class `BluetoothLeListAdapter` extending the `BaseAdapter`. The constructor of the adapter creates an empty `ArrayList` of `BluetoothDevice` class which will later be used to hold all the scanned devices.

```
public BluetoothLeListAdapter(Activity baseActivity) {       super();
mLeDevices = new ArrayList<BluetoothDevice>();       mInflator =
(LayoutInflater)
baseActivity.getSystemService(Context.LAYOUT_INFLATER_SERVICE);
```

```
}
```

In the `addDevice` method, `mLeDevices` are updated with newly scanned devices. This task will be achieved by:

```
public void addDevice(BluetoothDevice device) {       if
(!mLeDevices.contains(device)) {            mLeDevices.add(device);        }
}
```

For the `BaseAdapter` class, `getView()` method needs to be overridden to perform custom tasks when the view is updated. In our example, it includes inflating the view and updating the list:

```
@Override public View getView(int i, View view, ViewGroup viewGroup) {
ViewHolder viewHolder;       // General ListView optimization code.       if
(view == null) {            view = mInflator.inflate(R.layout.listitem_device,
null);          viewHolder = new ViewHolder();
viewHolder.deviceAddress = (TextView)
view.findViewById(R.id.device_address);          viewHolder.deviceName =
(TextView) view.findViewById(R.id.device_name);
view.setTag(viewHolder);       } else {          viewHolder = (ViewHolder)
view.getTag();       }       BluetoothDevice device = mLeDevices.get(i);
final String deviceName = device.getName();       if (deviceName != null &&
deviceName.length() > 0)          viewHolder.deviceName.setText(deviceName);
else          viewHolder.deviceName.setText("This device is unknown...");
viewHolder.deviceAddress.setText(device.getAddress());       return view;
}
```

`ViewHolder` is a simple class created to hold two `TextViews`. These will show the device name and device address respectively.

```
static class ViewHolder{       TextView deviceName;       TextView
deviceAddress;
}
```

The corresponding Bluetooth device from the array will be chosen and put in the view holder. If the device name is null, it will put a custom text in the view holder. To set the view holder in the view, you simply need to call `view.setTag(viewHolder)`

The final UI will look like this:

Android ListView containing all the Bluetooth devices in the proximity

The complete code for this application can be found in the resources of this chapter. Feel free to modify the code and use it in your own examples.

Android app - heart rate monitor application

In this section, we will give a complete example of a Central-Peripheral communication using Android. Peripheral is the device which advertises itself and Central responds to the advertisement and initiates the connection process. Normally, the peripheral is an IoT device which broadcast its identity and central is a cell phone which connects to the device to exchange data.

We have discussed various scenarios where a device can act as peripheral and the cellular phone can act as a central to start the communication process. Since we are not focusing on the hardware in this chapter, the complete cycle of Central-Peripheral communication is acquired by one of the software tools discussed in Chapter 2, *BLE Hardware, Software, and Debugging Tools*. We are using Nordic Connect Application to deploy a GATT server as a peripheral. We will configure the server with minimum efforts which will advertise built-in heart rate service.

Once the server will be deployed, we will move to the Android application which will act as a central. The application will explore the services and characteristics advertised by the server. It will also write the values to the peripheral.

Deploying the GATT server using Nordic Connect

In this section, we will learn how to deploy a GATT Server using Nordic Connect Android application. This server will behave as a peripheral and advertise itself. In order to start the server creation process, you need to first download and install the Nordic Connect Application. You can download Nordic Connect app using this link:
https://play.google.com/store/apps/details?id=no.nordicsemi.android.mcp:

1. The first step in deploying the server is to start the Nordic Connect application and acquire the following screen:

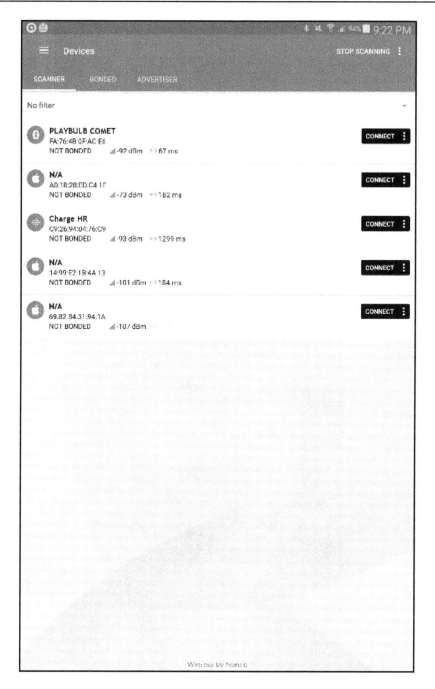

Nordic Connect main interface showing surrounding BLE devices

2. Select the menu bar on the left and select **Configure GATT Server:**

Nordic's menu bar showing configure GATT server option

3. You will see that the current status of the GATT server is disabled. Click on the disabled option on the top and click on plus icon. It will ask you to name your GATT configuration:

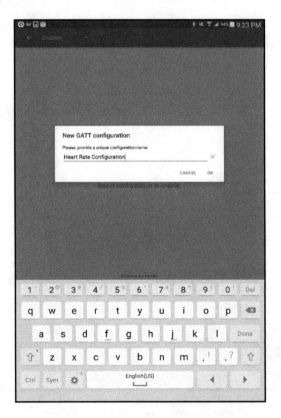

Nordic's option to name the configuration

4. Click on **Add Service** button appeared at the bottom of the configuration name. Upon clicking the customs service drop down, you can find a dialog showing pre-built services:

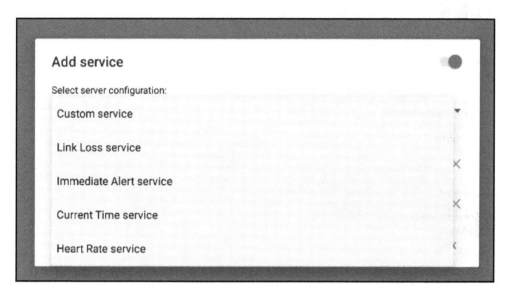

Nordic's pre-built services

5. Choose **Heart Rate service**. It will show the characteristics associated with the HR service. Do not forget to turn the service on by toggling the button in the top right corner of the dialog box.

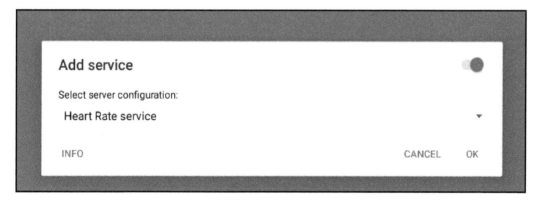

Toggle button to enable the service

6. Click **OK** and see the associated characteristics with the service. These characteristics are pre-built and will be used in the advertisements:

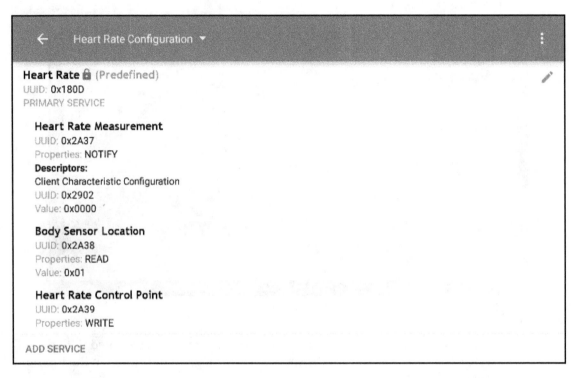

Characteristics view for the added HR service

7. Go back on the main screen and click **Advertiser**. Click on the plus sign on the bottom right corner of the screen. Fill the advertiser information and add records. These records are the advertising data that will be broadcasted. Select **Tx Power Level** and **Flags** as default advertisement value.

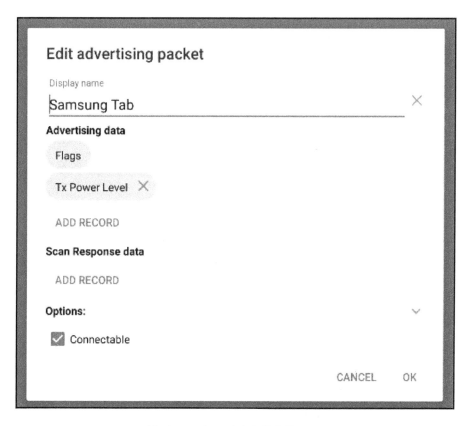

Advertisement packet creation in the Nordic Connect app

8. You can put additional data in the advertisement packet which doesn't require a connection. It is not recommended to put sensitive information in the advertisement packets. Once you are done, make sure that the advertiser and the server are enabled:

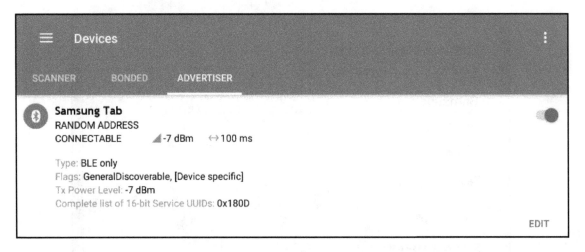

Advertiser is enabled in the Peripheral

Writing central-side Android apps

We will continue our Bluetooth Scanning application created in the previous section.

1. Upon configuring the server, the app should list the advertising peripheral in the list:

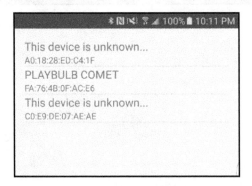

Scanner App is showing the configured server with the MAC C0:E9:DE:07:AE:AE

2. Since the activity is a `ListView`, we can implement `onListItemClick()` method to capture the clicks on the list of the Bluetooth devices:

```
@Override protected void onListItemClick(ListView l, View v,
int position, long id) {        final BluetoothDevice device =
mLeDeviceListAdapter.getDevice(position);
Toast.makeText(this, "Device: " +device.getName() + " and
address: " + device.getAddress(), Toast.LENGTH_SHORT).show();
Intent intent = new Intent(this,
DeviceDescriptionActivity.class);
intent.putExtra("Bluetooth_Device", device);
startActivity(intent);
}
```

This implementation will get the device according to the position clicked in the list and call the `DeviceDescriptionActivity` using `Intent`. The Bluetooth device is then passed to a new activity where it is handled for the further processing.

Refer to the Android API documentation to study the ways to send data from one activity to another using intent:
`https://developer.android.com/reference/android/content/Intent.h tml`.

3. In `DeviceDescriptionActivity`, bring the views that will show the information about the services. In our application, we have three different `textview` that will hold the MAC address, Service UUIDs, and Characteristics UUIDs respectively. The Bluetooth device is get using the `getParcelable()` method:

```
@Override protected void onCreate(Bundle savedInstanceState) {
super.onCreate(savedInstanceState);
setContentView(R.layout.activity_device_description);
BluetoothDevice device =
getIntent().getExtras().getParcelable("Bluetooth_Device");
descriptionMac = (TextView)
findViewById(R.id.device_description_mac);
descriptionServices = (TextView)
findViewById(R.id.device_description_services);
descriptionCharacteristic = (TextView)
findViewById(R.id.device_description_characteristics);
descriptionMac.setText(device.getAddress());
descriptionServices.setText("");
descriptionCharacteristic.setText("");        connect(device);
}
```

4. Once you get the device object using
   `getIntent().getExtras().getParcelable("Bluetooth_Device")`, it is
   easier to get the device address and populate the corresponding `textview`. Once
   the address view is populated, the `connect()` method is called with the
   Bluetooth device in the argument:

```
private BluetoothGatt mBluetoothGatt;
public void connects(BluetoothDevice device) {      if
(mBluetoothGatt == null) {         mBluetoothGatt =
device.connectGatt(DeviceDescriptionActivity.this, false,
mGattCallback);       }
}
```

5. The implementation of the `connect()` method is simple. You will call a
   `connectGatt` method on the Bluetooth device to get `BluetoothGatt` object.
   This object will be used to discover services and characteristics later. In the
   previous code snippet, you can see an object `mGattCallback`, which refers to a
   `BluetoothGattCallback` class. This class provides implementable methods like
   `onConnectionStateChange()`, `onServicesDiscovered()`,
   `onCharacteristicRead()`, `onCharacteristicChanged()` and
   `onCharacteristicWrite()`. The implementation of onConnectionStateChange
   in our application is given here:

```
public final BluetoothGattCallback mGattCallback = new
BluetoothGattCallback() {       @Override      public void
onConnectionStateChange(BluetoothGatt gatt, int status, int
newState) {         super.onConnectionStateChange(gatt, status,
newState);         switch (newState) {                  case
STATE_CONNECTED:                 Log.i("gattCallback",
"STATE_CONNECTED");                  gatt.discoverServices();
break;           case STATE_DISCONNECTED:
Log.e("gattCallback", "STATE_DISCONNECTED");
break;           case STATE_CONNECTING:
Log.i("gattCallback", "STATE_CONNECTING");
break;           default:
Log.e("gattCallback", "STATE_OTHER");              }
    }
```

6. Once the device is connected and the `mBluetoothGatt` object is collected. We will now call `gatt.discoverServices()` method to start the service discovery process. The optimal time to call this method is when the connection process is successful. In `onConnectionStateChange()` method, you can put `gatt.discoverServices()` in the `STATE_CONNECTED` switch-case. `STATE_CONNECTED` makes sure that the method is not called when the connection is in the process. This will result in an unsuccessful attempt of service discovery:

   - So far we have successfully connected to the peripheral device and started the service discovery phase
   - The next step is to capture all the discovered services. This functionality can be achieved by overriding an `onServicesDiscovered` method in the `mGattCallback` object

   ```
   @Override public void onServicesDiscovered(BluetoothGatt gatt,
   int status) {       super.onServicesDiscovered(gatt, status);
   final List<BluetoothGattService> services = gatt.getServices();
   for (int i = 0; i < services.size(); i++) {           final int
   localIndex = i;          runOnUiThread(new Runnable() {
   @Override              public void run() {
   descriptionServices.append(services.get(localIndex).getUuid().t
   oString() + "n");               }            });      }
   }
   ```

7. We will get a list of discovered services using `gatt.getServices()` and iterate through the list to populate the text view. The final output should look like this:

Discovered Services in the sample app

Each UUID in the previously mentioned picture corresponds to the services created on the GATT server using Nordic Connect App.

8. The next step is to capture the characteristics of Heart Rate Service. The UUID used for the Heart Rate service is:

```
private UUID heartRateServiceUUID =
UUID.fromString("0000180d-0000-1000-8000-00805f9b34fb");
```

9. In order to read the characteristic, we wanted to put an event-based characteristic reading. `buttonReadCharacteristic` was implemented in the `onCreate` method whose initial visibility was set to `INVISIBLE`. This ensures that the button will only appear when all the services are discovered. After discovering the services, the visibility of the button is set to `VISIBLE` by calling:

```
buttonReadCharacteristic.setVisibility(View.VISIBLE);
```

In Android, an easier way to connect a button to a method is by declaring a public method and passing a view in it. This method will then appear in the **Design** section of the button properties in the `layout.xml` file.

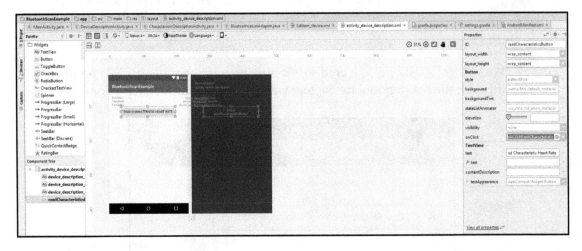

onClick property is appearing on the right hand side which button is in focus.

Define a method in your `DeviceDescriptionActivity` class by writing:

```
public void onClickReadCharacteristic(View button){        final
List<BluetoothGattCharacteristic> characteristics =
mBluetoothGatt.getService(heartRateServiceUUID).getCharacteristics();
descriptionServices.append("Reading Characteristics: n");        for (int i =
0; i < characteristics.size(); i++) {            final int localIndex = i;
runOnUiThread(new Runnable() {                @Override            public
void run() {
descriptionServices.append(characteristics.get(localIndex).getUuid().toStri
ng() + "n");            }            });        }
}
```

The list characteristics is obtained by calling the service with a specific UUID. Once the service is received, the `getCharacteristics()` method is called to obtain a list of `BluetoothGattCharacteristic`. This contains the list of all the characteristics offered by the Heart Rate service. We can now simply call a loop and append the characteristic UUIDs in a text view that we already created. The output on the Android screen should look like:

BluetoothScanExample

C0:E9:DE:07:AE:AE
00001801-0000-1000-8000-00805f9b34fb
00001800-0000-1000-8000-00805f9b34fb
0000180d-0000-1000-8000-00805f9b34fb
Reading Characteristics:
00002a37-0000-1000-8000-00805f9b34fb
00002a38-0000-1000-8000-00805f9b34fb
00002a39-0000-1000-8000-00805f9b34fb

READ CHARACTERISTIC HEART RATE

A sample application showing the characteristics of Heart Rate Service.

These UUIDs are the characteristics deployed on the server side on Nordic Connect app:

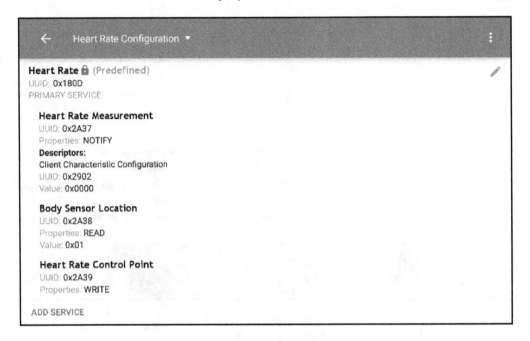

Heart Rate characteristic 0x2A37, 0x2A38 and 0x2A39 are matching the central output of UUIDs

The characteristic UUID xxxx2a37-xxxx-xxxx-xxxx-xxxxxxxxxxxx matches the peripheral characteristic **Heart Rate Measurement** with the UUID of 0x2A37. The characteristic UUID xxxx2a38-xxxx-xxxx-xxxx-xxxxxxxxxxxx matches the peripheral UUID of **Body Sensor Location** with the UUID of 0x2A38 and the characteristic UUID of xxxx2a39-xxxx-xxxx-xxxx-xxxxxxxxxxxx matches the peripheral characteristic of **Heart Rate Control Point** with the UUID of 0x2A39.

Writing data on the characteristic

The last topic of this section deals with writing the data on the characteristic. For writing process, the first thing to check is if the characteristic needs an enable notification request. If central does not explicitly enable the notifications on the peripheral, the peripheral will not send notifications back upon receiving the data. The notifications can be enabled by calling:

```
mBluetoothGatt.setCharacteristicNotification(characteristic, true);
```

The android application attached with the resource of this chapter does not perform writing to characteristic function but it can easily be extended by just implementing the following function in the application:

```
private boolean writeToCharacteristic(BluetoothGattCharacteristic
characteristic, byte[] aValue, boolean aWithResponse) {      boolean
isWritten = false;       if (characteristic != null) {
characteristic.setValue(aValue);
characteristic.setWriteType(aWithResponse ?
BluetoothGattCharacteristic.WRITE_TYPE_DEFAULT :
BluetoothGattCharacteristic.WRITE_TYPE_NO_RESPONSE);          isWritten =
mBluetoothGatt.writeCharacteristic(characteristic);      } else {
Log.e(TAG, "writeToCharacteristic() - Characteristic is null! " +
characteristic);      }      return isWritten;
}
```

This method takes the `BluetoothGattService`, `BluetoothGattCharacteristic`, value to write (a byte array) and a response flag (pass true if the response is required, false if not). The method can be called every time some write functionality is required. First, we will handle `NullPointerException` on the characteristic objects to make sure that the characteristic is eligible to write data on. After that, `characteristic.setValue()` method is called with the byte array to write data on the characteristic object. At this point, the data is not written on the characteristic. The second step is to define if the peripheral will return a response or not. This functionality can be achieved by writing `BluetoothGattCharacteristic.WRITE_TYPE_DEFAULT` or `BluetoothGattCharacteristic.WRITE_TYPE_NO_RESPONSE`. After defining the response type, we will not write the value to the peripheral. `mBluetoothGatt.writeCharacteristic()` method is called with the local characteristic object to fulfil the writing process.

If you have multiple values to write on different characteristics, it is a best practice to introduce delays between consecutive write requests. Normally peripherals are not fast processors and centrals overpower them by sending fast consecutive requests. The best way to handle this problem is to introduce `Thread.Sleep()` between each writes request.

The complete code of central-peripheral communication is given in the resources of this chapter.

Bluetooth Over-The-Air device firmware update

In this section, we will take a look at Nordic's device firmware update Android library that can be used to upload firmware on an nRF5x series based device. In nRF5x Series Nordic chips and S-series SoftDevices, you can perform OTA device firmware update. Consider, you have a device that features nRF5x SoC and you wrote an Android application to communicate with that device. Using OTA DFU library, you can pull the new firmware from the cloud to the app and wirelessly upgrade the firmware in your device. Refer to the following diagram to understand this flow:

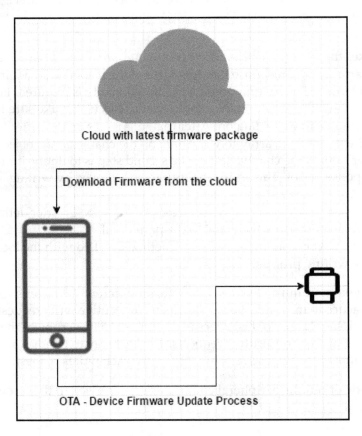

The latest version of Nordic's OTA DFU library is version 1.3.0 and can be downloaded from here `https://github.com/NordicSemiconductor/And roid-DFU-Library`. From Android library version 1.3.0, Nordic introduced a secure device firmware update mechanism. A detailed description of BLE secure DFU bootloader can found at `http://infocent er.nordicsemi.com/index.jsp?topic=%2Fcom.nordic.infocenter.sdk 5.v12.0.0%2Fble_sdk_app_dfu_bootloader.html&cp=4_0_0_4_3_1`.

Adding Android DFU library using Gradle

Make a new Android project and open `build.gradle` file in the `app` folder. In the `dependencies` section, add:

```
Compile 'com.nordicsemi.anroid:dfu:1.3.0'
```

A complete view of a simple `build.gradle` file in `app` module is given as:

```
apply plugin: 'com.android.application'
android {
    compileSdkVersion 23
    buildToolsVersion "23.0.3"
    defaultConfig {
        applicationId "com.test.firmwareapplication"
        minSdkVersion 23
        targetSdkVersion 23
        versionCode 1
        multiDexEnabled true
        versionName "1.0"
        testInstrumentationRunner
"android.support.test.runner.AndroidJUnitRunner"
    }
    buildTypes {
        release {
            minifyEnabled false
            proguardFiles getDefaultProguardFile('proguard-android.txt'),
'proguard-rules.pro'
        }
    }
}

dependencies {
    compile fileTree(dir: 'libs', include: ['*.jar'])
    compile 'com.android.support:appcompat-v7:23.3.0'
    compile 'no.nordicsemi.android:dfu:1.3.0'
    compile 'com.android.support:design:23.3.0'
```

```
        testCompile 'junit:junit:4.12'
    }
```

DFUService

In order to perform DFU, you will first write a `DFUService` that is extended by `DfuBaseService` provided by Nordic's API. In this class you will write your implementation of three significant methods:

- `getNotificationTarget()`, where you will define your activity in which you want to receive the notifications from the service.
- `onBind()`, where you will return the communication channel to the service.
- `isDebug()`, where you will define if your application is in debug or in production mode

```java
public class DfuService extends DfuBaseService {
    public DfuService() {

    }@Override
    protected Class<? extends Activity> getNotificationTarget() {
        return NotificationActivity.class;
    }
    @Override
    public IBinder onBind(Intent intent) {
        // TODO: Return the communication channel to the service.
        throw new UnsupportedOperationException("Not yet implemented");
    }
    @Override
    protected boolean isDebug() {
        return BuildConfig.DEBUG;
        }
    }
```

Here, the `NotificationActivity` is a simple activity defined as:

```java
public class NotificationActivity extends AppCompatActivity {
    @Override
    protected void onCreate(Bundle savedInstanceState) {
        super.onCreate(savedInstanceState);

        // If this activity is the root activity of the task, the app is
not running
        if (isTaskRoot()) {
            // Start the app before finishing
            final Intent intent = new Intent(this, MainActivity.class);
```

```
            intent.addFlags(Intent.FLAG_ACTIVITY_NEW_TASK);
            intent.putExtras(getIntent().getExtras()); // copy all extras
            startActivity(intent);
        }

        // Now finish, which will drop you to the activity at which you
were at the top of the task stack
        finish();
    }
}
```

FirmwareUpdateActivity

This is the activity which will require DfuService class to upload the firmware. You will first need to implement a file chooser which will be responsible to pick the secure package for firmware. This can be achieved by including following package to your FirmwareUpdateActivity:

```
private static final int FIRMWARE_SELECT_CODE = 0;
private void showFileChooser() {
    Intent intent = new Intent(Intent.ACTION_GET_CONTENT);
    intent.setType("application/zip");
    intent.addCategory(Intent.CATEGORY_OPENABLE);

    try {
    startActivityForResult(
    Intent.createChooser(intent, "Select a File to Upload"),
    FIRMWARE_SELECT_CODE);
    } catch (android.content.ActivityNotFoundException ex) {
    // Potentially direct the user to the Market with a Dialog
    Toast.makeText(this, "Please install a File Manager.",
    Toast.LENGTH_SHORT).show();
    }
}
```

FIRMWARE_SELECT_CODE is a unique code that we are using while calling the intent. It will help us to filter a callback from this intent. When we will come back to our application from the file chooser, we will use FIRMWARE_SELECT_CODE for identification. Your intent from the file chooser will come back to onActivityResult, where you will write:

```
@Override
protected void onActivityResult(int requestCode, int resultCode, Intent
data) {
    switch (requestCode) {
        case FIRMWARE_SELECT_CODE:
```

```
        if (resultCode == RESULT_OK) {
            firmwareUri = data.getData();
            this.grantUriPermission(this.getPackageName(), firmwareUri,
Intent.FLAG_GRANT_READ_URI_PERMISSION);
            }
        break;
        }
    super.onActivityResult(requestCode, resultCode, data);
}
```

The next step is to write the method which will use `firmwareUri` to upload the firmware to the device. It will require two other parameters:

- `deviceAddress`: MAC address of the device you want to upload the firmware
- `deviceName`: Name of the device you want to upload the firmware

Identify these two things using basic Bluetooth LE scan (as described in the previous sections) and call:

```
private void startSecureDfuProcess(String deviceAddress, String deviceName)
{
    final DfuServiceInitiator starter = new
DfuServiceInitiator(deviceAddress).setDeviceName(deviceName).setKeepBond(tr
ue);
    starter.setZip(firmwareUri);
    starter.start(this, DfuService.class);
}
```

`DfuServiceInitiator` will take the parameters such as `deviceAddress`, `deviceName` and `firmwareUri` and by calling `DfuServiceInitiator.start()`, you can start the secure device firmware update process. Notice that there is no message listener registered yet. So, in order to hear the messages from the device firmware update process you will need a `DfuProgressListener`. Defined a global variable `mDfuProgressListener`:

```
private DfuProgressListener mDfuProgressListener = new
DfuProgressListener() {

    //Override methods here...
}
```

DfuProgressListener provides many methods to override certain behaviors. You can find following methods in it:

- onDeviceConnecting, is called when DfuService is making a connection to the device
- onDeviceConnected, is called when a connection is established
- onDfuProcessStarting, is called upon starting the Dfu process
- onDfuProcessStarted, is called when the Dfu process has started
- onFirmwareValidating, is called to indicate that service is validating the firmware package
- onDeviceDisconnecting, is called when the device is disconnecting from the service
- onDfuCompleted, is called when the DFU process if completed
- onDfuAborted, is called when the DFU process is aborted to some reason
- onError, is called if an error occurs in the DFU upload process

These hooks can be used to update the UI on the go. It will keep the user informed of the current processes. Once you implement these methods, you can register the mDfuProgressListener to the DfuService by calling in onResume method of your activity:

```
DfuServiceListenerHelper.registerProgressListener(this,
mDfuProgressListener);
```

By implementing this project, you will be able to successfully execute device firmware service and hear the message callbacks. The code is generic and can be executed from any activity. Just make sure that you have given location, Bluetooth and Uri read permissions to the activity.

Summary

In this chapter, we discussed a theoretical overview of BLE central and peripheral communication with an introduction of Android. We then moved towards a more practical approach of making an Android application that performs scanning functionality. We then discussed another example that showed a complete heart rate application that performs central-peripheral communication, including GATT connect, service discovery, characteristic discovery, characteristic read, characteristic write and notifications, and so on. In the next chapter, we will explore the world of Bluetooth Low Energy Beacons, its classification, implementation, and deployment.

4
Bluetooth Low Energy Beacons

Any modern technology can be well understood if we predict what it will become in the next 10 years. Let's journey back in time, when Bluetooth classic was launched and became available for customers. A key factor in its success was that it was available for everyone. The cellular phone manufacturers of that time were integrating this technology into their phones, which helped Bluetooth spread all over the world in no time. Bluetooth SIG predicted that the next era will contain IoT to its core and will be dominated by the IoT and that is very important to provide a low-power-consumption communication solution for devices. Thus, Proximity Beacon API was created as version 4.0 of the standard. In the previous chapters, we discussed various theoretical and practical concepts of Bluetooth Low Energy, and this chapter is no different. Here, we will discuss a new application of Bluetooth Low Energy—**BLE beacons**. This chapter contains the following topics:

- Introduction to Bluetooth Low Energy beacon technology
- Classification of BLE beacons—Eddystone and iBeacon
- Introduction to the Estimote SDK and Nearables
- Physical web with Eddystone using Estimote

Introduction to Bluetooth Low Energy Beacons

Bluetooth Low Energy Beacons are Bluetooth devices that broadcast their **universally unique identifiers** (**UUIDs**) in proximity. They are low-power devices attached to some physical object (such as a wall or an electronic device), which detects the nearby device (typically a cellular phone) and triggers some behavior. These behaviors normally contain a location-based task or a unique broadcast that can be preprogrammed on the beacons before deployment. Beacons assume that each person has a Bluetooth Low Energy device already listening to beacon calls and that the broadcasts will be received by them. In this chapter, we will use the Estimote as our primary Bluetooth beacon; it looks like this:

In this chapter, we used Long Range Location Beacons by Estimote. They are ideal for indoor locations and IoT prototyping. They come with built-in sensors (ambient light sensors and barometer sensor) and transmit up to eight simultaneous packets. You can order these beacons by going to `https://order.estimote.com/buy/location-devkit`.

Applications of Bluetooth Low Energy Beacons

In the modern era, technology seeks devices that are low maintenance and do not require effort to fulfill people's needs. Bluetooth Low Energy Beacons fulfill these requirements. Theoretically, Bluetooth Low Energy Beacons can be deployed anyplace where there is no restriction on transmitting Bluetooth signals. These beacons come with a built-in battery that can last for several years. They are independent devices with no need for maintenance, which makes them flexible to deploy anywhere.

With these inherent qualities, Bluetooth Low Energy Beacons come with a diverse application set. This section will discuss those applications to provide an overview of what can be achieved through this technology. You can then explore your imagination to come up with your own ideas to perform extraordinary tasks. The section will run from the easiest to the most complicated implementation of Bluetooth Low Energy Beacons.

Beginner applications

Stickers are probably the most widely used beacon application. They are tiny beacons that can stick to any surface (mostly valuable items). For example, imagine a beacon that sticks to your keys and constantly keeps track of its location. This beacon will warn you if you forget the keys in a restaurant. In this scenario, the sticker made the keys aware of their surroundings, which is the main purpose of the beacon technology. The sticker is in constant contact with the smartphone in order to know its position relative to the person. The same logic will be applied in our next example, but in a completely different scenario. Imagine you go shopping and, while passing through a clothing store, you get a beep on your cell phone. This is a Bluetooth Low Energy beacon sending a proximity broadcast regarding special offers. The application is not only beneficial for the customer but also for the store owner. If a beacon is put on every item, due to the built-in accelerometer, the beacon can now tell the owner how many times an item in the store is being picked up. This is valuable information and can be used by the store owner to study customer preferences.

Advanced applications

After understanding the basic beacon example, we can now move forward to understanding more complex applications. A proximity beacon can be put on the front door of a house. If the resident comes inside, it can turn on the lights and switch the radio to a particular channel. If the beacon has the ability to connect to a smart home hub, it can even control multiple things at once. For example, by knowing that the resident is in the vicinity, a Bluetooth beacon can perform customized tasks, such as playing the voicemail on the home phone or brewing coffee. The applications of beacons are not confined to smart homes. They can also be used to monitor a hand sanitizer dispenser in a hospital, for example, to make sure the employees are using it properly.

Indoor location is another significant milestone achieved by the Bluetooth beacon technology. We all use location technology when we commute or search for our favorite restaurant down the road, but before beacon technology, there was no easy way to track indoor location. Since GPS technology does not work indoors, Bluetooth beacons were introduced to enable indoor navigation. The Location Beacons work on a **triangulation algorithm** to calculate the precise location of a device. We will take a complete tutorial on how to implement indoor location in the next chapter.

Here's what configuring a purple beacon for indoor navigation looks like:

Beacon protocols

Just like any other technology, Bluetooth Low Energy Beacons work with a protocol set. It is like a profile that tells how a beacon is broadcasting information. This information is then handled by a device (typically a cell phone) with the same protocol. Both ends should know the definition of the broadcast data if they want a successful transmission. This is why the broadcast data type is defined by the type of protocol the beacon is using. In some cases, it broadcasts a UUID-based URI, while in other cases, it broadcast URLs or telemetry packets.

The protocol gives a definition to the data broadcast.

The broadcast data type is defined by the type of protocol the beacon is using. In some cases, it broadcasts a UUID-based URI, while in other cases, it broadcast URLs or telemetry packets. In this section, we will discuss two widely used beacon protocols: Eddystone by Google and iBeacon by Apple.

This protocol standard is not to be confused with the Bluetooth Low Energy protocols, which define the very fundamentals of BLE communications.

Google Eddystone

Google Eddystone is an open source beacon protocol by Google, compatible with Android and iOS. It comes with three kinds of data packets:

- Eddystone-UID
- Eddystone-URL
- Eddystone-TLM

In Google Eddystone, the process of setting up the beacon is called **provisioning** and can be achieved by an Android/iOS app provided by Google.

These apps can be downloaded from the following links:
Android: `https://play.google.com/store/apps/details?id=com.goog le.android.apps.location.beacon.beacontools`
iOS: `https://itunes.apple.com/us/app/beacon-tools/id1094371356`.

Once an object or a place is identified to deploy the beacon on or at, the next step is to choose the type of data the beacon will be transmitting. It can be either of the three types mentioned previously.

After selecting the data type, the third step is to register the beacon with the Google **Proximity Beacon API**. It is a cloud service offered by Google to let owners manage the data associated with a beacon using a REST interface. The interface enables real-time data monitoring and management remotely and allows Google products to react to the beacon. We will learn about the Proximity Beacon API in the upcoming sections.

Eddystone-UID

Eddystone-UID is a datatype supported by Eddystone beacons. It is a broadcast frame consisting of a 16-byte beacon ID, which contains a 10-byte namespace and a 6-byte instance. The 10-byte namespace is used to identify a group of Eddystone beacons, while the instance is used to identify a particular beacon. Both parts of the UID ensure the uniqueness of the beacon or set of beacons. The data structure of Eddystone-UID is as follows:

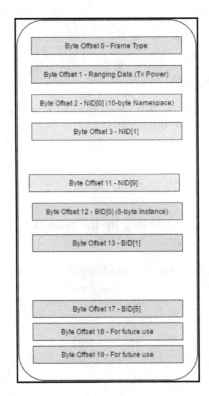

The default Eddystone-UID value is `EDD1EBEAC04E5DEFA017`. Eddystone-UID has an additional functionality to attach telemetry data. This telemetry data can be a URL for purposes of fleet management. Eddystone-UID with telemetry data is referred to as Eddystone-TLM.

Eddystone-EID

Eddystone-EID is another datatype Google Eddystone supports in their beacons. The frame broadcasts an encrypted **ephemeral identifier** (**EID**) that changes periodically. The rate can be described at the time of beacon registration with a web service. This broadcast identifier can then be resolved remotely by the same service it was registered to. This feature is added in the beacon for security applications, where the data needs secure transmission. For a man-in-the-middle attacker, the frame data will appear to be changing randomly, but at the receiver's end (with the correct web service), it is always resolvable. The frame type for Eddystone-EID is as follows:

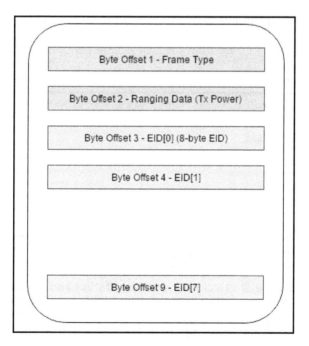

Eddystone-URL

The Eddystone-URL data format is the backbone of Google's physical web. The frame broadcasts a Google-encoded URL, which enables it to broadcast more information in fewer characters. Any device that has access to the Internet will be able to decode the URL and open the underlying web page. An example of a Google-encoded URL is `https://goo.gl/Aq18zF`.

The frame structure of Eddystone-URL is as follows:

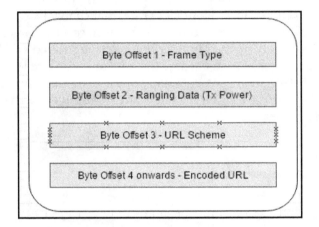

 The source code of Google Eddystone is available at `https://github.com/google/eddystone`.

Apple iBeacon

Just like Eddystone, Apple iBeacon is a beacon platform used for enabling location awareness for devices. The iOS device can then identify itself when it has entered beacon proximity. Unlike Eddystone, it is a closed implementation and native only to iOS users. The advertising packet for iBeacon consists of these four components:

- **UUID**: Universally unique identifier
- **Major**: Two-byte string to identify a group of beacons; similar to the namespace in Eddystone

- **Minor**: Two-byte string to identify individual beacons; similar to an instance in Eddystone
- **Tx Power**: Transmission Power

In iOS 7, the location was determined by the geographical location (outdoors only), known as **geolocation**. In iBeacon, this task is achieved by the identifier, which has become a part of the **Core Location API**. The location can now be tagged to a particular object. This region is now movable and can perform a location-based task for the current location. For example, an iBeacon can be put on a food truck that is mobile, and you will receive special offers whenever you come close to the food truck.

 An example iBeacon transmission frame looks like this:
`fb3b77a1-8328-44 cd-913a-82a122bc8513` Major 1 Minor 2. Here, `fb3b77a1-8328-44 cd-913a-82a122bc8513` is the UUID, `1` is the Major and `2` is the Minor.

	Compatibility	Profile	Data types	Ease of use	Security
Apple iBeacon	Android and iOS-compatible but native to iOS only	Proprietary protocol	UUID with Major and Minor	Simple to implement	No secure datatype like EID in Eddystone
Google Eddystone	Android and iOS-compatible, supported by any BLE beacon	Open source protocol	UID, EID, URL, and TLM	Simple to implement but requires coding in complicated scenarios	Data security using EIDs

Estimote beacons

Estimote Inc. was founded by Jakub Krzych and Lukasz Kostka in 2012. The company focuses on making low-power IoT devices that can stick to walls or objects. They use many technologies, such as Bluetooth Low Energy, Wi-Fi, HDMI, and mesh networking. Their devices normally come with sensors, such as motion, temperature, ambient light, magnetometer, pressure, and telemetry. They specialize in Bluetooth Low Energy Beacons with full support for Eddystone and iBeacon platforms. In this chapter, we will take a look at their SDK and practical examples that you can implement using Estimote beacons.

Estimote has a wide variety of beacons for different kinds of applications. I recommend buying Long Range Location Beacons because they provide much more flexibility and options. You can buy Estimote beacons at `http ://estimote.com/#get-beacons-section`.

Estimote Bluetooth Low Energy Beacons feature a 32-bit ARM Cortex CPU with a 2.4 GHz radio for Bluetooth communication. The battery typically lasts for 1-5 years, depending on the type of beacon you are using. The beacon is fully compatible with Android and iOS devices. The typical range of an Estimote beacon is up to 70 meters (over 200 feet). In this section, we will look at Estimote examples for iBeacon and Eddystone. Here's a view of Eddystone beacons:

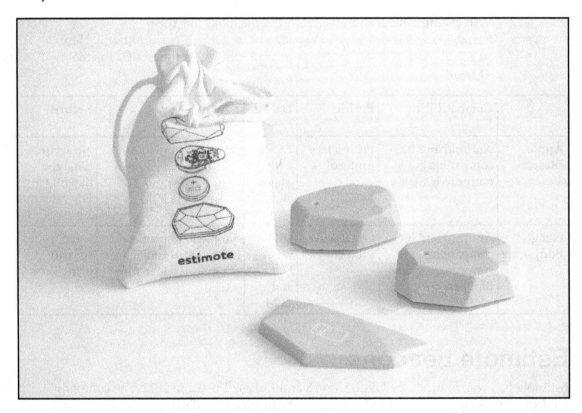

Estimote SDK for Android

Estimote comes with its own software development kit to enable developers to implement their own projects. In this section, we will discuss the Estimote Android SDK and how to implement your customized projects. The Estimote SDK works for Android 4.3+ and requires devices that support Bluetooth Low Energy. So make sure that before starting development, you have a newer device that supports Bluetooth Low Energy. I'm also assuming that you have basic Android development knowledge to start implementing the projects. For any query regarding Android development, consult `https://developer.andr oid.com/index.html`.

In a nutshell, the Estimote Android SDK provides following features to the user:

- Beacon scanning (scan the surrounding for beacons)
- Beacon monitoring (monitor entry and exit of devices in a region)
- Nearable discovery (used for Estimote Stickers)
- Eddystone scanning
- Reading and writing characteristics for Estimote proximity and Location Beacons
- Analytics

Getting started

Adding the Estimote SDK to your Android project is simple. Add the following code to your project's Gradle script (Module: app):

```
compile 'com.estimote:sdk:0.11.1@aar'
```

As mentioned in the previous chapter, Android SDK version 23 or greater requires the explicit `ACCESS_COARSE_LOCATION` permission at runtime. I'm giving you the code snippet for these permissions; for details, you can consult the *Android Bluetooth Low Energy* section of the previous chapter.

Paste the following code to `AndroidManifest.xml`:

```
<uses-permission android:name="android.permission.ACCESS_COARSE_LOCATION"
/>
```

Use the following code in `Activity.java`:

```
// --------------- MANUAL PERMISSIONS----------------
String[] permissionString = new
String[]{Manifest.permission.ACCESS_COARSE_LOCATION,
Manifest.permission.ACCESS_FINE_LOCATION};
```

```
yourActivity.requestPermissions(permissionString,
yourPermissionRequestCode);
// --------------------- CATCHING PERMISSIONS -------------------------
@Override
public void onRequestPermissionsResult(int requestCode, String
permissions[], int[] grantResults){
 if(requestCode == yourPermissionRequestCode)
 {
 ... //Do something based on the user click.
 }
}
```

Although Estimote cannot automatically ask for the ACCESS_COARSE_LOCATION permission on your behalf, it does provide a method to make things simpler and check whether all the requirements have been met before we start making any Estimote-related calls. This functionality can be achieved by calling this method in your onResume() method:

```
SystemRequirementsChecker.checkWithDefaultDialogs(this);
```

This method uses default dialog boxes to ask user permissions, such as the Bluetooth or ACCESS_COARSE_LOCATION permissions.

Background monitoring and ranging

Background monitoring in the Estimote SDK is used to trigger an event whenever the device enters or leaves the range of the beacon. Consider an example of a designer store. The user gets a **Welcome to our designer store** notification on his cell phone as soon as he walks in the store. The detection of the application in the proximity of the beacon is achieved by background monitoring.

Background monitoring can detect the presence of your application near an Estimote beacon, but to make it truly powerful, one must use *ranging*. Monitoring only enables the app to recognize whether the user has entered or exited an area, while ranging, on the other hand, causes an application to constantly scan for nearby beacons. Let's consider the previous example to understand these two terms. Monitoring will detect as you enter the designer store, but if your application runs a ranging task, it will receive all the broadcasts from nearby beacons with their UIDs and Major/Minor values. With ranging, you can give special promotions to the customer.

Another important use of the ranging feature in the application is proximity estimation. It means estimating your location with respect to the beacon. A beacon broadcasts with a specific signal strength, which gets weaker the farther it travels. Proximity estimation uses this to recognize how far the application is from the beacon. If the signal strength is strong, it is closer to the beacon. Otherwise, it is farther away. There are four regions described by the Estimote API:

- **Immediate**: This indicates strong signals, up to a few centimeters
- **Near**: This indicates medium signals, up to a few meters
- **Far**: This indicates weak signals, of more than a few meters
- **Unknown**: This is when the signals are very weak

To study how to apply ranging in an Android application, we will assume that we have an Estimote beacon with a UUID of `7a2fc4ba-03a3-4241-8a32-eaa2e471218a`. Add the following code to the `onCreate()` method of your Android application:

```
BeaconManager beaconManager = new BeaconManager(this);
Region region = new Region("Entry Region",
            UUID.fromString("7a2fc4ba-03a3-4241-8a32-eaa2e471218a"), null,
null);
```

`BeaconManager` is the class used to interact with Estimote beacons. It provides various functionalities, such as:

- Monitoring and ranging the Estimote beacon
- Discovering Nearables
- Scanning nearby Eddystone beacons

The region, on the other hand, defines a location around the beacon. It is defined by the UUID and Major and Minor value of the beacon. In the previous code, we created a `BeaconManager` object to do all kinds of communication with the beacon, and a `region` object, which will be required to form a virtual region around the beacon (using the corresponding UUID).

Let's start with the monitoring implementation. Considering the objects created previously, we will call a `beaconManager.connect` method with a new `ServiceReadyCallback()` object as an argument. This object will implement the `onServiceReady()` method, which is called once the service is ready to perform `beaconManager` tasks:

```
beaconManager.connect(new BeaconManager.ServiceReadyCallback() {
    @Override
public void onServiceReady() {
        beaconManager.startMonitoring(new Region(
```

```
"D&G Store Region",
                UUID.fromString("7a2fc4ba-03a3-4241-8a32-eaa2e471218a"),
11435, 87936));
}
});
```

In this `beaconManager.connect` code, we are starting the monitoring over a region defined by the UUID of the beacon, `7a2fc4ba-03a3-4241-8a32-eaa2e471218a`, and its Major and Minor value (`11435` and `87936`, respectively). Now, the application is searching for this particular UUID with its Major and Minor values. We need to now implement the listener, which is called when the user walks into or leaves this region:

```
beaconManager.setMonitoringListener(new BeaconManager.MonitoringListener()
{
    @Override
public void onEnteredRegion(Region region, List<Beacon> list) {
        // Do something on entering a region
}
    @Override
public void onExitedRegion(Region region) {
// Do something on exiting a region
}
});
```

In this code, you can perform a particular task (typically a UI update) upon entering a particular region. In the case of our designer store example, we can notify the user when he enters or leaves the store.

 If multiple beacons are detected while monitoring a region, there will only be one entry and exit event each. It is important to put beacons in the order of their monitoring.

We will now connect the beacon manager to a beacon service and start the ranging operation:

```
beaconManager.connect(new BeaconManager.ServiceReadyCallback() {
        @Override
public void onServiceReady() {
            beaconManager.startRanging(region);
}
});
```

The onServiceReady() method is called when the BeaconService is ready to use. We will start the ranging process after the service is ready to be used by calling beaconManager.startRanging operation on the created beacon. At this point, the ranging process is successfully initiated. We can stop the ranging process by calling beaconManager.stopRanding(region) method. Normally the ranging start process should be called in the onResume() method and ranging stop process should be called at an onPause method in your activity.

The next step is to read the beacon broadcast and get the data out of it. As you know from the previous section, a beacon transmits in the form of a UUID and Major and Minor data packets. These Major and Minor data packets are not understandable unless we provide a definition to them. Let's continue with the previous store example.

The beacon at the door of the designer store is broadcasting its UUID and Major and Minor values. Let's assume that it is giving us a list of all the products, arranged according to their distance from the door. This whole list will be understood from the Major and Minor values. The definition of those Major and Minor values should be somewhere in the server/cloud in order to remotely change them. For now, assume the list of products in the beacon is the following:

```
private static final Map<String, List<String>> PRODUCTS_BY_BEACONS;

// TODO: replace "<major>:<minor>" strings to match your own beacons.
static {
    Map<String, List<String>> productsByBeacons = new HashMap<>();
productsByBeacons.put("11435:87936", new ArrayList<String>() {{
add("D&G Sweater for Men");
add("D&G Tuxedo");
add("D&G Girls Payjama");
}});
PRODUCTS_BY_BEACONS = Collections.unmodifiableMap(productsByBeacons);
```

This HashMap is a supposed piece of server data that provides the definition to the Major and Minor value of a broadcast. Let's now look at the implementation of this in the Android application:

```
private List<String> findProductsNearBeacon(Beacon beacon) {
    String beaconMajorAndMinorKey = beacon.getMajor() + ":" +
beacon.getMinor();
if (PRODUCTS_BY_BEACONS.containsKey(beaconMajorAndMinorKey)) {
return PRODUCTS_BY_BEACONS.get(beaconMajorAndMinorKey);
}
return null;
}
```

In this method, we are taking a `beacon` object in the argument. This is the scanned beacon that is broadcasting the Major and Minor values with its UUID. We are then concatenating these values using a : separator. The definition (ideally supplied by the server) is then responsible for giving the definition of this `<Major>:<Minor>` value. This definition is the list of all the products placed in the store with respect to their distance from the door.

One thing that magically appeared in the code is the `beacon` object. This object is the result of the beacon ranging listener, which can be implemented as follows:

```
beaconManager.setRangingListener(new BeaconManager.RangingListener() {
    @Override
public void onBeaconsDiscovered(Region region, List<Beacon> list) {
        // perform the operation on the list of sorted beacon in terms of
their distance from the phone.
    }
});
```

A `RangingListener` object is passed to the `beaconManager.setRangingListener` method after overriding the `onBeaconsDiscovered` method. This method will be called when all the beacons in the proximity are scanned. We will call our `findProductsNearBeacon()` method in the `onBeaconsDiscovered` method:

```
beaconManager.setRangingListener(new BeaconManager.RangingListener() {
    @Override
public void onBeaconsDiscovered(Region region, List<Beacon> list) {
if (!list.isEmpty()) {
            for(Beacon beacon: list){
                List<String> products = findProductsNearBeacon(beacon);
// update the UI here
                Log.d("D&G Store Beacon: " + beacon.getMacAddress(),
"Product Found: " + products);
            }
    }
    }
});
```

This method will go through all the beacons found in the proximity, find products near each beacon, and log the list of devices.

Estimote SDK for iOS

In this section, we will learn how to implement the same examples using iOS. The Estimote SDK is written in Objective-C, but it works perfectly with Swift as well. You need to provide a bridging header by following these steps:

1. Choose **New File** by right-clicking on the project's group.
2. Pick a **Header File** from the **iOS-Source** section.
3. Save the header file as `ObjCBridge.h`.
4. Import the Estimote SDK by calling `#import <EstimoteSDK/EstimoteSDK.h>` in the newly created file.
5. Go to **Build Settings**.
6. Find the **Objective-C Bridging Header** setting, and set it to `${PROJECT_NAME}/ObjCBridge.h`.

Once you are all set up with the bridging header, you can start using the SDK in Swift.

For iOS, we will use the same procedure and example as of the Android application. The first step is to make the `beaconManager` object, which will be responsible for all the communication between the beacon and the application. In your application function, in the `App:Delegate` class, write this code:

```
let beaconManager = ESTBeaconManager() //a global instance
self.beaconManager.delegate = self
```

As described before, iBeacon is a part of iOS's **Core Location service**; you need to explicitly ask the user for permission to access their location data. This can be done by going to the `Info.plist` file and adding a new row with the key `NSLocationAlwaysUsageDescription` and String type.

 The message for the permission can be anything polite.

When the code executes for the first time, it will show the message described. The next step is to request the authorization:

```
self.beaconManager.requestAlwaysAuthorization()
```

At this point, your application has the `beaconManager` object and the authorization to communicate to the beacon. The next step is to start the monitoring of the beacon. We will use the same UUID that we used in the previous example:

```
self.beaconManager.startMonitoringForRegion(CLBeaconRegion(
    proximityUUID: NSUUID(UUIDString: "7a2fc4ba-03a3-4241-8a32-
eaa2e471218a")!,
    major: 123, minor: 123, identifier: "D&G Store Region"))
```

We are telling the `beaconManager` to start monitoring for the region with `7a2fc4ba-03a3-4241-8a32-eaa2e471218a` beacon. The Major and Minor values used here are both `123`. Once we start monitoring, we need to explicitly call `stopMonitoringForRegion`; otherwise, it will keep on monitoring until the app is uninstalled.

Go ahead and add the following implement in your `AppDelegate`. This implementation is called the `didEnterStore` delegate method:

```
func beaconManager(manager: Any, didEnterStore region: CLBeaconRegion) {
let storeNotification = UILocalNotification()
storeNotification.alertBody =
"You have just entered a D&G Store. " +
"Let us know if we can help you with anything."
UIApplication.sharedApplication().presentLocalNotificationNow(storeNotifica
tion)
}
```

This book assumes you already know the basics of Android and iOS development. We need to request the notification permission from the user, which can be achieved with this:

```
UIApplication.sharedApplication().registerUserNotificationSettings(
UIUserNotificationSettings(forTypes: .Alert, categories: nil))
```

As of now, we have successfully implemented a background monitoring feature in the iOS application. The next step is to implement the ranging. Keeping in mind the previous example Android application, we will implement the ranging example. Go to the `ViewController` implementation file and set up a second beacon manager. We will also have a beacon region property to hold the virtual region surrounding the beacon:

```
// Adding the beacon manager and the beacon region
let beaconManager = ESTBeaconManager()
let beaconRegion = CLBeaconRegion(
        proximityUUID: NSUUID(UUIDString: "7a2fc4ba-03a3-4241-8a32-
eaa2e471218a")!,
        identifier: "D&G Store Region")
```

```
override func viewDidLoad() {
super.viewDidLoad()
self.beaconManager.delegate = self
self.beaconManager.requestAlwaysAuthorization()
}
```

The `viewDidLoad` method will self-delegate the `beaconManager` function and request the authorization for the location. After asking for the authorization, we will go again to the `ViewController` class and write this:

```
override func viewWillAppear(animated: Bool) {
super.viewWillAppear(animated)
self.beaconManager.startRangingBeaconsInRegion(self.beaconRegion)
}

override func viewDidDisappear(animated: Bool) {
super.viewDidDisappear(animated)
self.beaconManager.stopRangingBeaconsInRegion(self.beaconRegion)
}
```

This will allow the application to start and stop the ranging process on a `beaconRegion` object. Keeping the same *server* data in mind, which provides the definition to the Major and Minor value of the beacon, we will now implement the `productsNearBeacon` method:

```
func productsNearBeacon(beacon: CLBeacon) -> [String]? {
let beaconKey = "(beacon.major):(beacon.minor)"
if let products = self.productsByBeacons[beaconKey] {
let sortedProducts = Array(products).sorted { $0.1 < $1.1 }.map { $0.0 }
return sortedProducts
}
return nil
}
```

When the beacon with a Major and Minor value of `11435:87936` will appear, it will grab the list of all the products. This list is already sorted with respect to the distance of products from the beacon.

The `productsNearBeacon()` method can be called by the ranging delegate, simply by calling the following function:

```
func beaconManager(manager: Any, didRangeBeacons beacons: [CLBeacon],
in region: CLBeaconRegion) {
if let nearestBeacon = beacons.first, places =
productsNearBeacon(nearestBeacon) {
// TODO: update the UI here
println(places) // TODO: remove after implementing the UI
}
}
```

Unlike the Android example, we are calling the `productsNearBeacon` method only on the nearest beacon. The list is already sorted, and by grabbing the first object in the list (`beacons.first`), it ensures the nearest beacon comes first.

Physical web using Estimote

Estimote supports both Eddystone and Apple's iBeacon protocol, but to enable physical web on an Estimote beacon, Eddystone is enough. You will require a proximity beacon that broadcasts at least one packet at a time. It can either transmit an iBeacon packet or an Eddystone packet. Make sure to enable only one type before configuring it.

In order to switch your beacon to Eddystone, follow these steps:

1. Go to `https://play.google.com/store/apps/details?id=com.estimote.apps.main` to download the Estimote Android application.
2. Sign up if you don't already hold an account with Estimote.
3. Tap on the beacon you want to switch to Eddystone.
4. Choose **Packets**.
5. Enable **Eddystone-URL** with the URL you want it to broadcast.

The beacon is now transmitting the physical web URL you just configured. In order to receive the physical web broadcasts from the beacon on your Android phone, go to **Google Chrome** | **Settings** | **Privacy** | **Enable Physical Web**. Provide the location permission if not done already.

Download the Google physical web application from Play Store by following this link: `https://play.google.com/store/apps/details?id=physical_web.org.physicalweb`

Once the URL is received on the phone, the user will then click on it, which should automatically take it to a website. An example here is a smart parking meter. It is broadcasting a URL to the nearby users, which will take them to a website where the user can pay for space automatically. This is a typical implementation of the physical web as described by Google. I invite you to think of more examples where the concept of physical web is implemented.

Estimote cloud interface

Estimote provides a cloud interface, where you can register your beacons and perform some analytics. It also provides short links to developers resources, forums, and the community portal. The main interface looks like this:

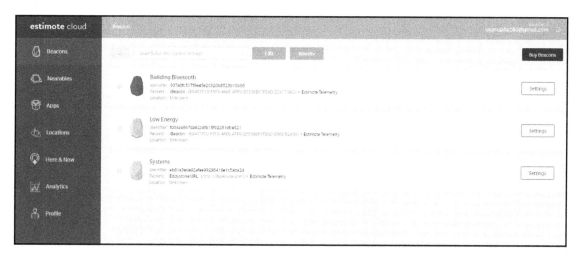

The cloud interface provides control over the settings of Bluetooth beacons. It lets you configure iBeacon and various Eddystone data packets of. It also let you call the Estimote Location service. Although it gives you the flexibility to write to the beacon, you need the Estimote application to apply these settings. Sign up for an Estimote account and start using Estimote cloud at `https://cloud.estimote.com/`.

A graphical view of how to apply settings in the Estimote Android application is given here:

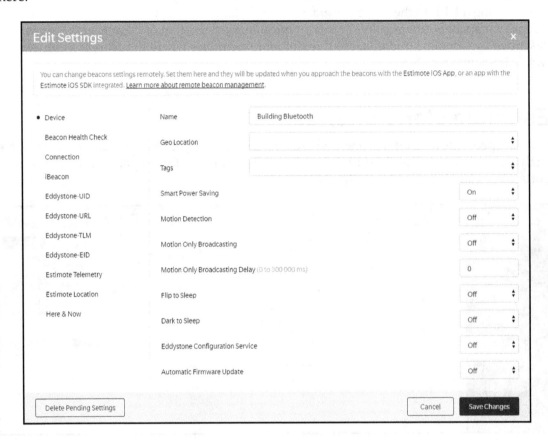

Summary

In this chapter, we explored Bluetooth Low Energy Beacons and discussed their applications with real-life scenarios. We also discussed BLE protocols such as Google Eddystone and Apple iBeacon, two of the most widely used protocols in beacon technology. The chapter later discussed a practical application built on Android and iOS using Estimote beacons. The discussion continued with a simple implementation of a physical web-based URL broadcast using Estimote. Finally, we discussed the Estimote cloud interface to view and control your beacons using the cloud. In the next chapter, we will move forward and discuss an even more complicated example of Bluetooth indoor location using Estimote beacons.

5
BLE Indoor Navigation Using Estimote Beacons

Bluetooth technology was first made for small-distance data transfer, but we are now pushing the limits by implementing it in various scenarios. Indoor navigation is one of the scenarios where Bluetooth has made its mark. Since its early adoption in cellular phones, it has been steadily progressing at an exponential rate. But in order to keep up with information technology, this technology had to fashion itself with the support of the Internet of Things. Thus, the Bluetooth SIG decided to revise the very architecture of Bluetooth (in version 4.0), replacing it with a low-power, simpler, and stronger version. One of the advantages of this Bluetooth Low Energy architecture is that it can be used for many simple yet practical implementations. In this chapter, we will discuss one such implementation, indoor navigation.

These are the topics we will cover:

- Estimote Location Beacons and triangulation algorithm
- Estimote Location SDK on iOS
- Implementing indoor navigation

Introducing indoor navigation

In 2005, Google launched a satellite-imagery-based system for outdoor navigation, and in the next 11 years, Google Maps evolved to include real-time traffic conditions, panoramic street views, and route planning. For the first time, people were introduced to a new system of navigation where they were not dependent on physical maps or transit booklets. They could confidently go to a new city and navigate like a local.

Here is a Google Maps car that collects street view data:

The practicality of Google's idea fits well with the concept of the Internet of Things. In the IoT's context, either a user needs to know the location of the IoT devices or vice versa. To understand this, consider the example of Levi's Stadium in Santa Clara, United States. It is a football stadium in the San Francisco Bay Area and has served as the home of the San Francisco 49ers in the NFL since 2014. When they were building the stadium, they needed to install an indoor navigation system. Since this is a state-of-the-art stadium, people should have a smart way to find the closest hot dog stand, get their food delivered right to their seats, or find the closest washroom. This problem could not be solved by Google Maps because it relies on satellite navigation, which does not work indoors.

Although it posed some security threats, the indoor navigation problem in Levi's Stadium was addressed using Wi-Fi technology. The problem would be much easier to solve using Bluetooth Low Energy beacons, but the technology was not available at the time of construction. The working indoor navigation in the stadium can be experienced using the Levi's Stadium app on Google Play (`https://play.google.com/store/apps/de tails?id=com.venuenext.stadium`) and on iOS (`https://itunes.apple .com/ca/app/levis-stadium/id891480007`).

In light of this example, you can think of hundreds of situations where indoor navigation is required. Estimote, in their BLE Location Beacons, addressed this issue and provided a BLE-based indoor navigation solution. In this chapter, we will study how indoor navigation can be easily implemented in our homes or any other indoor location.

Estimote Location Beacons and triangulation

In this section, we will study how triangulation algorithms can be helpful in determining indoor location.

 Although Estimote Beacons don't claim to use a triangulation algorithm in their **data science**, it is one of a few possibilities.

Estimote Location Beacons

In the previous chapter, we had a thorough introduction to Estimote Beacons and how they can be implemented in various applications. In this chapter, we will especially target Estimote Location Beacons and implement an indoor navigation system. Estimote Location Beacons are custom-tailored for location services and can be purchased from their website. These beacons are good for prototyping and enterprise applications. They also support mesh networking and have built-in sensors (temperature, motion, light, and magnetometer) and broadcast up to eight simultaneous packets.

 In order to set up these beacons for basic configuration, refer to the previous chapter.

The easy integration of Estimote Location beacons with Estimote Nearables enables us to search tagged devices in the vicinity. They get highlighted on a map with their locations, like this:

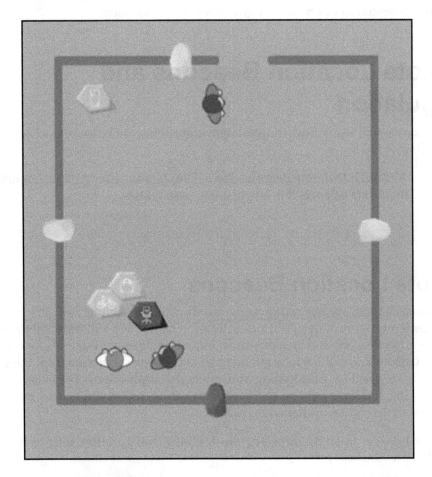

This diagram represents a room where beacons on the walls are Location Beacons, which help in navigation, while the Nearable tags in the middle of the room represent the devices present in the vicinity. These devices can be anything that you can stick the Estimote tag on. The interface can be used in many practical scenarios; for example, a shop owner can monitor the on-display items. It can also be used to monitor an office space to see whether some office supplies or important/confidential items have been moved away from a room.

Triangulation for indoor navigation

Estimote is hesitant to disclose their indoor navigation algorithm, but from the looks of it, a triangulation algorithm makes the most sense. In this chapter, we will learn how a triangulation algorithm can be used for indoor navigation. The topic is necessary because if you know the theory behind the working model of indoor navigation, you can come up with your own implementation of the algorithm and build your very own Bluetooth system.

According to Estimote, indoor location is based on beacons and advanced data science for establishing the position of users and objects.

Triangulation is a concept of trigonometry and geometry. It is a process of estimating the location of a point by forming triangles to it from the location of known things. We will take a simple room example to explain how triangulation can work in an indoor location. In order to implement triangulation in this room, imagine that each wall has a Bluetooth beacon dedicated to it (as Estimote recommends):

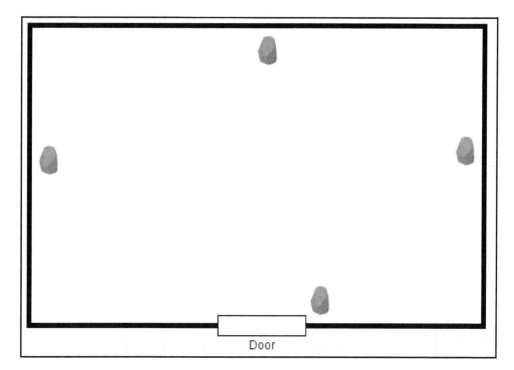

Door

The triangulation will work best if the room is triangular, but since that is not the case, it is recommended you use one beacon for each wall. Now imagine the signals from a beacon, which are omnidirectional, and a cellphone representing the user, for a setup such the following:

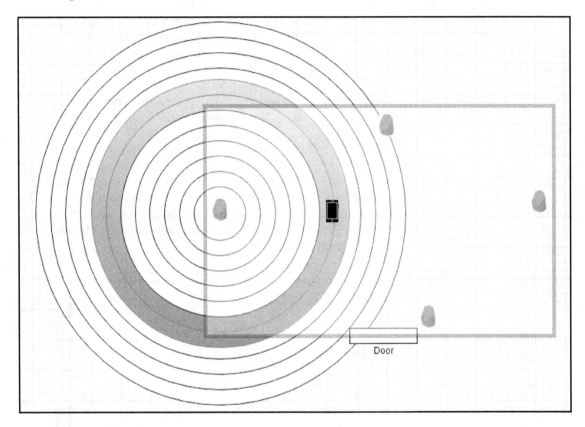

Door

Based on the signal strength, the user can be anywhere in the purple circles. So it is not possible to know the location of the user based on just one beacon. But now imagine a scenario where two beacons are doing the same thing:

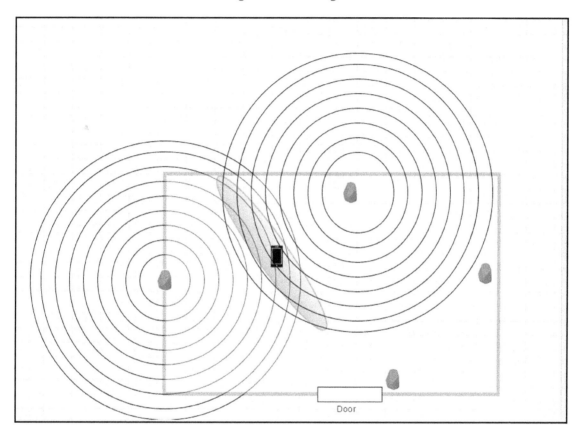

Since now we are receiving signal strength data from two beacons, we can roughly estimate that the location of the user is somewhere in the blue ellipse. The location is not yet successfully determined, so we will take the approximations from the other two beacons too:

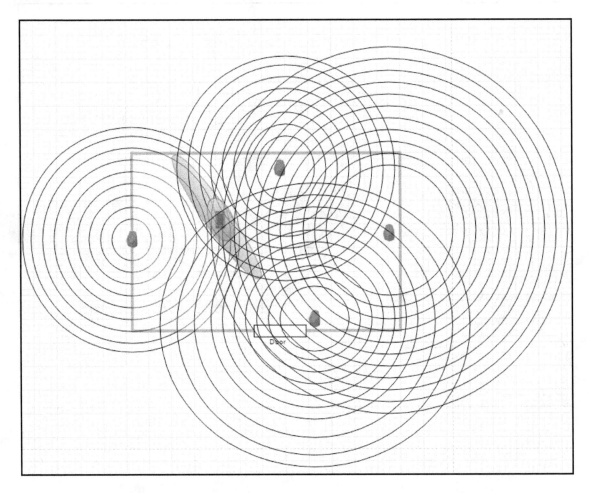

By using all the four beacons in the room, we can approximate a very reasonable location (red ellipse) of the user. By adding beacons, we will increase the accuracy, which is what Estimote recommends as well.

The same method is used to find the approximate location of an earthquake's epicenter and the presence of a user in a cellular network. Using triangulation, you can implement your own indoor navigation. Another useful method of determining approximate location is called **trilateration** (working with distances instead of angles), but it is less accurate indoors. The best-case scenario is if we combine the two methods to refine the location.

 GPS signals works on the trilateration concept, where the distance is determined using speed and time.

Indoor location using the Estimote Location Beacons

Let's first set up indoor location using the iOS app.

Setting up an indoor location using the iOS app

In this section, we will learn to set up basic indoor navigation using Estimote Location Beacons. The simple implementation contains two major components:

- At least four Estimote Location Beacons
- The Estimote iOS app (https://itunes.apple.com/ca/app/estimote-indoor-location/id963704810)

Once you download the application, you can register the beacons on Estimote Cloud (as described in the previous section). The next step is to put the beacons on each wall of the room (preferably vertically, at chest height, and in the middle of the wall). The next step is to start the calibration process in the iOS app. In the calibration process, you will start by placing the phone next to a beacon and letting the app recognize the location. Then walk along the walls of the room and keep on placing the cellphone next to each beacon to register its location. The process is complete when you complete the walk and return to the first registered beacon.

The app will now upload the map to Estimote Cloud for further development. It will then show a map of your room and update your position in real time:

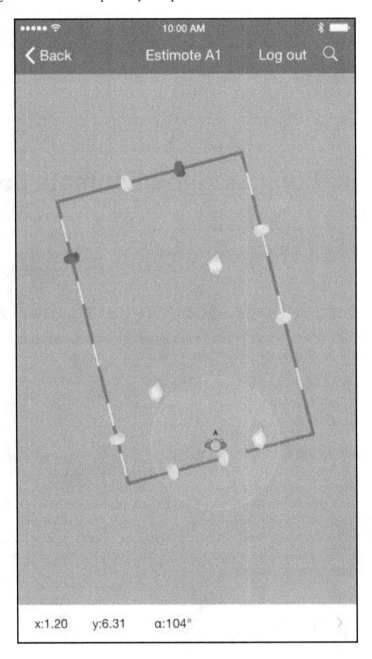

Estimote Location SDK on Swift

In this section, we will study the functionalities offered by the Estimote Location API. The section will cover the most important classes and methods you will need to implement your own examples. For a detailed implementation, refer to the next section.

EILIndoorLocationManager

`EILIndoorLocationManager` is the core class you will use to make your interaction with Estimote Cloud. The important methods of this manager class are:

- `startMonitoringForLocation`: This method will tell you whether you are inside or outside a monitored location. The location is an `EILLocation` object provided by the Estimote API.
- `stopMonitoringForLocation`: This method will stop the monitoring process for a location.
- `startPositionUpdatesForLocation`: This method starts the process to deliver position updates for a monitored location.

EILLocation

The `EILLocation` class represents the location of a Bluetooth beacon for an indoor environment. The instances of this object will come up when you call the `EILRequestFetchLocation` class. Some important properties of this class are:

- **Geographical location**: As described in Estimote Cloud
- **Latitude**: Latitude of the location
- **Longitude**: Longitude of the location
- **Orientation**: Orientation to magnetic north in degrees (clockwise)

EILLocationBuilder

`EILLocationBuilder` is a `builder` object for creating various `EILLocation` objects. You can programmatically create a new location by specifying the shape, orientation, and details of the location (walls, doors, and so on). Some important methods for this class are:

- `setLocationBoundaryPoints`: Specify the boundary points of the location

- `setLocationOrientation`: Set the orientation of the room with respect to magnetic north
- `addBeaconWithIdentifier:withPosition`: Place a beacon in a location
- `addDoorsWithLength`: Place a door on a boundary segment
- `addWindowWithLength`: Place a window on a boundary segment

EILOrientedPoint

The `EILOrientedPoint` class represents a geometric point with the orientation. Some important properties and methods of this class are:

- `Orientation`: Defines the orientation of the point
- `X`: Represents the x coordinate of the point
- `Y`: Represents the y coordinate of the point
- initWithX:y:orientation: Lets you create a point with given x, y, and orientation values

EILPositionedBeacon

The `EILPositionedBeacon` class provides information about the beacon. The important properties and methods are:

- `position`: Position of the oriented beacon
- `major`: Major value of the beacon
- `minor`: Minor value of the beacon
- `initWithBeaconIdentifier`: Initializes a new beacon object by position, color, proximity, UUID, or Major or Minor values

EILRequestAddLocation

The `EILPositionedBeacon` class is used to save a new location to Estimote Cloud. The two important methods of this class are:

- `initWithLocation`: Gets a new object for a request
- `sendRequestWithCompletion`: Sends the request object (`EILRequestAddLocationBlock`)

EILRequestFetchLocation

`EILRequestFetchLocation` is used to fetch a saved location from Estimote Cloud. The two important methods of this class are:

- `initWithLocationIdentifier`: Requests a new object for a request by its identifier
- `sendRequestWithCompletion`: Sends the fetch request to Estimote Cloud

EILRequestModifyLocation and EILRequestRemoveLocation

The `EILRequestModifyLocation` and `EILRequestRemoveLocation` classes represent the modification and removal of a location in Estimote Cloud. These methods are similar to `EILRequestFetchLocation`. The `sendRequestWithCompletion` method is used to request the Estimote Cloud to perform the action.

 In this section, we studied various important classes, methods, and properties of the Estimote SDK for iOS. To get complete information about the SDK, visit `http://estimote.github.io/iOS-Indoor-SDK/`.

Implementing indoor navigation using the Estimote Location SDK on Swift

The Estimote Indoor Location SDK is only available for iOS devices. Although it is not available for Android, it is possible to write your own SDK around Estimote Beacons that can use the Beacon parameters to approximate the location. Writing your own SDK is out of the context of book, which is why we will only focus on the Estimote SDK for iOS. In order to start with it, you need the following things:

- Mac machine with Xcode 7
- iPhone 4S (or newer)
- Estimote account
- At least four registered Estimote Location Beacons (refer to previous chapters for registering Estimote Beacons)

Once you fulfill all the prerequisites, make sure that you have already set up your room on Estimote Cloud (as described in the previous section). It is time to jump into the development environment and set up a project. Make a **Single View Application** in Xcode and name the project accordingly. For our purposes, we are calling the project BLESystemsIndoor. Once you make a new project, go to the terminal and install the SDK with the following:

```
cd {PATH_BLESSytemsIndoor}
{PATH}$ sudo gem install cocoapods
$ echo -e 'target "BLESystemsIndoor" donpod "EstimoteIndoorSDK"nend' >
Podfile
$ pod install
```

Close Xcode and reopen your project using the .xcworkspace file.

You can also go to https://github.com/Estimote/iOS-Indoor-SDK#installationand manually install the repository.

I recommend using CocoaPods as it is much easier to download and link the repository.

The execution of pod install command should look like this:

```
Usamas-MacBook-Pro:BLESystemsIndoor usamaaftab$ echo -e 'target "BLESystemsIndoor" do\npod "EstimoteInd
oorSDK"\nend'>Podfile
Usamas-MacBook-Pro:BLESystemsIndoor usamaaftab$ pod install
Setting up CocoaPods master repo
  $ /usr/bin/git clone https://github.com/CocoaPods/Specs.git master --progress
  Cloning into 'master'...
  remote: Counting objects: 1058358, done.
  remote: Compressing objects: 100% (29/29), done.
  Receiving objects:  80% (852651/1058358), 230.76 MiB | 1.31 MiB/s
```

You will see the following message once it complete the process:

```
Usamas-MacBook-Pro:BLESystemsIndoor usamaaftab$ echo -e 'target "BLESystemsIndoor" do\npod "EstimoteInd
oorSDK"\nend'>Podfile
Usamas-MacBook-Pro:BLESystemsIndoor usamaaftab$ pod install
Setting up CocoaPods master repo
  $ /usr/bin/git clone https://github.com/CocoaPods/Specs.git master --progress
  Cloning into 'master'...
  remote: Counting objects: 1058358, done.
  remote: Compressing objects: 100% (29/29), done.
  remote: Total 1058358 (delta 6), reused 2 (delta 2), pack-reused 1058325
  Receiving objects: 100% (1058358/1058358), 372.91 MiB | 1.43 MiB/s, done.
  Resolving deltas: 100% (491002/491002), done.
  Checking connectivity... done.
  Checking out files: 100% (134425/134425), done.
Setup completed
Analyzing dependencies
Downloading dependencies
Installing EstimoteIndoorSDK (2.3.2)
Installing EstimoteSDK (4.13.1)
Generating Pods project
Integrating client project

[!] Please close any current Xcode sessions and use `BLESystemsIndoor.xcworkspace` for this project fro
m now on.
Sending stats
Pod installation complete! There is 1 dependency from the Podfile and 2 total pods installed.
Usamas-MacBook-Pro:BLESystemsIndoor usamaaftab$
```

The Estimote SDK is written in Objective-C, but it can be used with Swift; you need to bridge the header by following these steps:

1. Go to the **Project Navigator** window and right-click on the project group.
2. Create a header file and save it as `ObjCBridge.h`.
3. Write this line to import the SDK: `#import "EILIndoorSDK.h"`.
4. Go to **Build Settings** in your project.
5. Find the **Objective-C Bridging Header** setting and set it to the newly created `ObjCBridge.h` file.

Once you are done making the bridge between the SDK's Objective-C and your current language, Swift, you can move ahead and get your **App ID** and **App Token** by registering your application on Estimote Cloud. In order to do this, log in to your Estimote Cloud account and click on **Apps**:

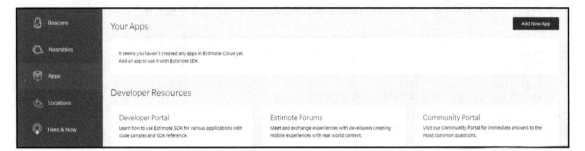

Click on your app in order to get the **App ID** and **App Token**:

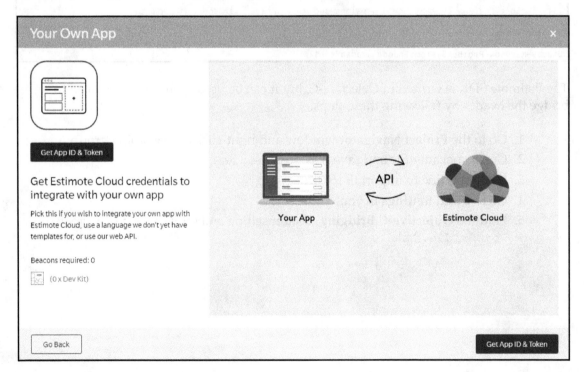

Provide the app name and app description and get the **App ID** and **App Token**:

Your access credentials to Estimote Cloud API:

App ID: blesystemsindoor-2hj

App Token: 789cd5d8ad58fe3771a1de1d033a3cd2

Now that you have your **App ID** and **App Token**, you can return to your project and add the indoor location manager to your view controller:

```
import UIKit

class ViewController: UIViewController, EILIndoorLocationManagerDelegate {

  let locationManager = EILIndoorLocationManager()

  override func viewDidLoad() {
  super.viewDidLoad()
  self.locationManager.delegate = self
  }

  override func didReceiveMemoryWarning(){
  super.didReceiveMemoryWarning()
  }
}
```

This `locationManager` is the base manager for the Estimote SDK, and you will interact with it many times during your development. Now link the Estimote Cloud **App ID** and **App Token** by replacing `<App ID>` and `<App Token>` with you own ID and token:

```
ESTConfig.setupAppID("<App ID>", andAppToken: "<App Token>")
```

The next step is to create a `location` object and get the location from the cloud:

```
var location: EILLocation! //a class level object to hold the location
```

Ask for the location using the following code:

```
let fetchLocationRequest = EILRequestFetchLocation(locationIdentifier: "my-
kitchen")
fetchLocationRequest.sendRequestWithCompletion { (location, error) in
 if location != nil {
 self.location = location!
 } else {
 print("can't fetch location: (error)")
 }
}
```

The `EILRequestFetchLocation` class is used to request the location from Estimote Cloud for the currently authorized user. The `my-kitchen` parameter is the identifier of your location. You will find this identifier in Estimote Cloud under the **Locations** tab. The next step is to start the position updates. Within the `viewDidLoad()` method, add the following code:

```
self.locationManager.startPositionUpdatesForLocation(self.location)
```

The `startPositionUpdatesForLocation()` method is to start the delivery of position updates from the specified location (in our case, `my-kitchen`). At this point, we have started the position updates, but we still need to provide some delegate methods to receive the coordinates. In order to do this, use the following code:

```
func indoorLocationManager(manager: EILIndoorLocationManager!,
  didFailToUpdatePositionWithError error: NSError!) {
  print("failed to update position: (error)")
}

func indoorLocationManager(manager: EILIndoorLocationManager!,
  didUpdatePosition position: EILOrientedPoint!,
  withAccuracy positionAccuracy: EILPositionAccuracy,
  inLocation location: EILLocation!) {
  var accuracy: String!
  switch positionAccuracy {
  case .VeryHigh: accuracy = "+/- 1.00m"
  case .High: accuracy = "+/- 1.62m"
  case .Medium: accuracy = "+/- 2.62m"
  case .Low: accuracy = "+/- 4.24m"
  case .VeryLow: accuracy = "+/- ? :-("
  case .Unknown: accuracy = "unknown"
  }
  print(String(format: "x: %4.2f, y: %4.2f, orientation: %4.0f, accuracy:
%@",
  position.x, position.y, position.orientation, accuracy))
}
```

Here, we implemented two methods with the same name (remember polymorphism). The `indoorLocationManager(manager: EILIndoorLocationManager!, didUpdatePosition position: EILOrientedPoint!, withAccuracy positionAccuracy: EILPositionAccuracy, inLocation location: EILLocation!)` function is where the updated location comes in. Let's understand what we just did.

We created a variable named `accuracy`, which will hold the string we want to show for accuracy. It will change according to the accuracy coming from the `EILPositionAccuracy` object from the SDK. EILPositionAccuracy describes the accuracy of the position determined by the SDK. In order to visualize the accuracy, consider a circle around a given location. A bigger circle will result in worse accuracy. The string values defined under the accuracy string variable are from Estimote; you can use custom values according to your needs.

The location updates are coming under the object position of the EILOrientedPoint type . EILOrientedPoint represents a geometric position with orientation. In order to extract the coordinates from this object, simply call position.x, position.y, or position.orientation.

By doing this, you will be able to get the real-time position of the user. At this stage, your code should look like this:

```
import UIKit

class ViewController: UIViewController, EILIndoorLocationManagerDelegate {

  let locationManager = EILIndoorLocationManager()
  var localLocation: EILLocation!

  override func viewDidLoad() {
  super.viewDidLoad()

  self.locationManager.delegate = self
  ESTConfig.setupAppID("blesystemsindoor-2hj", andAppToken:
"789cd5d8ad58fe3771a1de1d033a3cd2")

  let fetchLocationRequest = EILRequestFetchLocation(locationIdentifier:
"my-kitchen")
  fetchLocationRequest.sendRequestWithCompletion{ (location, error) in
  if location != nil{
  self.localLocation = location!
  }else{
  print("can't fetch location: (error)")
  }

  }

  self.locationManager.startPositionUpdatesForLocation(self.localLocation)
  }

  override func didReceiveMemoryWarning() {
  super.didReceiveMemoryWarning()
  // Dispose of any resources that can be recreated.
  }
```

```swift
func indoorLocationManager(manager: EILIndoorLocationManager,
didFailToUpdatePositionWithError error: NSError) {
print("unable to fetch the update position: (error)")
}

func indoorLocationManager(manager: EILIndoorLocationManager,
didUpdatePosition position: EILOrientedPoint, withAccuracy
positionAccuracy: EILPositionAccuracy, inLocation location: EILLocation) {

var accuracy: String!
switch positionAccuracy{
case .VeryHigh: accuracy = "+/- 1.00m"
case .High: accuracy = "+/- 1.62m"
case .Medium: accuracy = "+/- 2.62m"
case .Low: accuracy = "+/- 4.24m"
case .VeryLow: accuracy = "+/- ?"
case .Unknown: accuracy = "unknown"
}

print(String(format: "x: %4.2f, y: %4.2f, orientation: %4.0f, accurage:
%@", position.x, position.y, position.orientation, accuracy))
}
}
```

We have real-time coordinates for indoor navigation, the next step is to trigger some event based on your location. In order to do this we will keep track of the coordinates of the user and trigger an event if a user reach to a particular position (in our case, the main door of the house is where we want to trigger some event). Within your `indoorLocationManager()` method, write this code:

```swift
let frontDoor = EILPoint(x: 8.1, y: 3.2)
if position.distanceToPoint(frontDoor) < 2 {
  // Do some action: Why not trigger a welcome message for the guests?
}
```

You can implement hundreds of ideas just by using this example. Refer to the previous section, in order to gain more knowledge about the SDK. You will need it when you implement your own projects.

Summary

In this chapter, we had an overview of how to use the Estimote SDK using SWIFT. The API works in collaboration with Estimote Cloud and can be used to request and push data. Using the information given in this chapter, you can easily implement your own apps. As of now, Estimote only supports iOS and can only be used by iPhone 4S+, but soon, it will be available for Android as well. In the coming chapter, we will study the theoretical and practical aspects of Bluetooth Mesh technology with the help of CSRMesh.

6

Bluetooth Mesh Technology

In the age of electronics, the Internet emerged as a global system to connect every computer system around the globe. It revolutionized communication processes and forced us all to move from circuit-switch networks to packet-switch networks. The basic definition of the Internet states that it is a network of networks, where each node follows the TCP/IP model to communicate with another node. The idea of connecting everything is appealing, but in which topology type should we do it? Should it be bus, star, ring, daisy chain, or point-to-point?

Let's discuss this problem on a smaller scale. A rather private version of the Internet is called an **intranet**, which is a private network accessible to only a group of people. It can be an office network, hospital network, school network, and so on. If we have 10 computers in an intranet, there should be a way to connect those computers so that they can communicate with each other. Generally, a WLAN is connected in a star topology where a central hub (Wi-Fi router) offers connection to all the computers/cellular phones and creates a star topology.

Here are some other topologies:

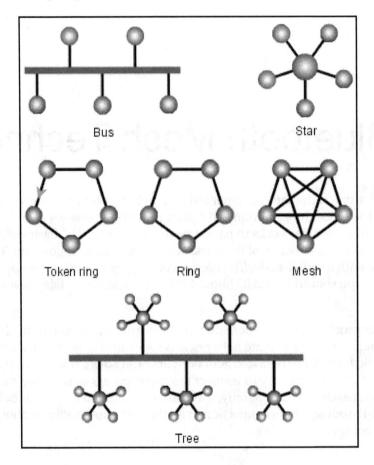

Just like a WLAN, traditional Bluetooth also followed a star-based topology. Let's take an example of a cell phone that acts as a hub and connects a lot of devices:

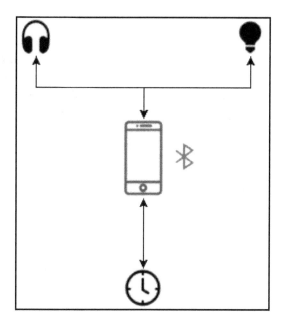

In this chapter, we will explore another significant topology type found in computer networks and see how that topology fits into Bluetooth Low Energy. We will discuss the following:

- Mesh topology and its implementation in ZigBee and Bluetooth Low Energy
- Routing and flooding algorithms in a Bluetooth Low Energy mesh
- Security complications in a Bluetooth Low Energy mesh
- The CSRMesh control application on Android

Introduction to mesh networking

Self-configuring networks in which the nodes are connected in a mesh topology are called **mesh networks**. In a mesh topology, each node acts as a router to transmit traffic between peers if the destination node is not directly connected to the source node. This formation of nodes is reliable because each node is connected to other nodes, providing alternate routes to any possible incoming packet. These networks are also self organized, so if a node leaves the network, the network will reconfigure itself for potential new routes for forwarding packets. The change of topology will occur when a node leaves or joins a mesh network. This phenomenon can happen rapidly in a situation where nodes are highly mobile.

In order to understand the basic mesh mechanism, consider three wireless devices: **Device 1**, **Device 2**, and **Device 3**. As shown in the following diagram, **Device 1** is not in the range of **Device 3** but is in the range of **Device 2**. On the other hand, **Device 3** is in the range of **Device 2** as well. In this situation, if **Device 1** wants to transfer packets to **Device 3**, **Device 2** will act as a *router* to route the packets. The topology will look something like this:

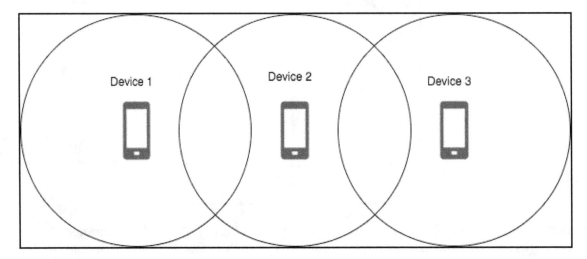

There are many kinds of mesh networks present in traditional computer networking:

- **Wireless mesh networks (WMNs)**
- **Mobile ad hoc networks (MANETs)**
- **Wireless sensor networks (WSNs)**

Advantages of mesh networking

Mesh networking brings many advantages to the communication mechanism of a network. These advantages are also applicable to Bluetooth Low Energy communication, so it is good to have a look at it. Mesh networking promises that there is no central point of routing; each node fulfills its responsibility of routing the packets across the network. This division provides many routes to a single destination. So in theory, there is always a way to route packets from destination *A* to destination *B* until at least one route is completed.

In order to understand this point in detail, consider a mesh network:

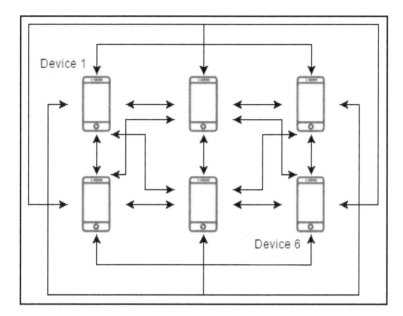

This is a fully connected mesh where every node has a direct connection to every other node in the mesh. In this scenario, even if **Device 1** is unable to use four of its routes, it will still be able to connect to **Device 6**. The situation is only possible because there is always some route available and the network is not dependent on a single point of routing (like traditional WLAN networking). This brings high stability to the mesh network.

Another significant advantage of mesh networking over a traditional star topology is that every additional node brings more stability to the network instead of being an overhead. If more nodes join a mesh network, there will potentially be more routes available to the network. Hence, the routing algorithm fulfills an important role in this. Another advantage linked to this is that every additional node brings with it additional capacity to handle a high volume of data flow.

A mesh network is easily expandable as well. Since nodes are self configured, as soon as they join a network, they start routing the packets. This helps in applications where easy expandability is required.

Mesh networking in wireless technologies

Theoretically, the traditional mesh topology is applicable to any kind of communication network, but it is mostly used in wireless technologies. If a new node is joining, it can join the network without any need for a physical wired connection, which makes it perfect for fast-changing networks (for example, vehicular networks). The implementation of this topology is limitless, but as of this chapter, we will see how it is useful for:

- Wireless mesh networks
- ZigBee
- Bluetooth Low Energy

Wireless mesh networks

WMNs were one of the earliest adoptions of mesh technology. Alongside Wireless Ad-hoc Networks, they established a very practical foundation of mesh networking. By successfully implementing routing algorithms, WMNs were able to implement the technology on an industrial level.

WMNs have a similar topology to mobile ad hoc networks, but they show slow or no mobility. In WMNs, a change in the topology occurs when a host leaves the network. WMNs also do not show tremendous traffic abnormalities because mesh networks are normally used under controlled circumstances. WMNs consist of different nodes with distinct working characteristics. The classification is as follows:

- **Wireless mesh clients** are typically the mesh nodes, such as laptops, computers, tablets, mobiles, and wearable tech devices
- **Wireless mesh routers** are the middle point between mesh clients and mesh gateways and route the traffic in a WMN
- **Wireless mesh gateways** are used to communicate with another WMN or with the Internet

Since they are an introductory topic, WMNs are not discussed in detail here. Let's move forward to the implementation of mesh networks in ZigBee.

Mesh networking in ZigBee

In the world of IoT, ZigBee was probably the first low-power technology that adopted a mesh topology. To give you an introduction about ZigBee's technology, it is a low-power, low-cost wireless standard designed for monitoring purposes. It is built on IEEE standard 802.15.4 and is normally used in a personal network environment. The typical range of ZigBee in a pure line of sight is somewhere around 10-100 meters, but the inherent advantage in the mesh topology gives ZigBee an extendable range. Although it is a low-power device, the use of mesh networks has given it strong adoption in the industry. In a typical ZigBee mesh network, there are three kinds of devices:

- **Coordinators** are the most powerful kind of devices in a ZigBee network. A coordinator is the root of the tree that makes a ZigBee mesh. It can communicate with other meshes as well and store information about the network, such as route information and security keys.
- **Routers** are the devices that relay packets through a ZigBee network.
- **End Devices** are the low-powered, low-cost devices that have enough functionality to talk to their parent routers.

A typical ZigBee network looks something like this:

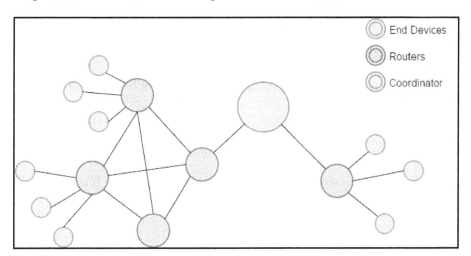

ZigBee **End Devices** do not make up a mesh; rather, the mesh networking is achieved on the router level. As you can see in the diagram, **Routers** are creating a mesh while **End Devices** are gaining indirect access to the whole mesh. Technically, it is neither purely mesh nor star topology, but it combines the usefulness of both in a small environment.

The reason ZigBee failed to gain popularity in consumer electronics is its absence from cellular devices. Even though it comes with the advantages of a mesh topology, a normal user is unable to communicate with it. Another reason ZigBee failed to gain wide popularity is because it doesn't come with strong security and is prone to breaches. Having said that, industries did use this technology in a wider concept of wireless sensor networks for over 20 years.

Mesh networking in Bluetooth Low Energy

Mesh networking has never been a part of the Bluetooth SIG until their release of Bluetooth Low Energy. The SIG reshaped the whole idea of Bluetooth communication in version 4.0. The low-energy version of Bluetooth is now more device friendly, as it offers low power and low cost. Although Bluetooth Low Energy v4.0, v4.1, and v4.2 provided a foundation for the Internet of Things, they missed their mark by not providing two things:

- High speeds with longer range
- Mesh networking capabilities

These two complaints were rectified in their announcement of Bluetooth 5. It is not called Bluetooth v5 or v5.0, but Bluetooth 5. Although the Bluetooth SIG disclosed their plans to launch Bluetooth 5 at the Bluetooth World Conference of 2016 (in Santa Clara), it was not sure when exactly they were going to do it. In June 2016, in a media event in London, the SIG unveiled Bluetooth 5 with a special focus on the Internet of Things.

Steve Hegenderfer (director of developer programs in Bluetooth SIG) described his excitement about the technology in a blog post on November 22, 2016 (`https://blog.bluetooth.com/author/shegenderfer`).

Traditionally, Bluetooth technology was born in a star-based topology, where the devices were unable to communicate with each other in a mesh. This was the fundamental problem with Bluetooth before it was widely adopted in IoT. Fortunately, Bluetooth 5 was unveiled last year, which brings:

- 2x speed as compared to Bluetooth 4.2
- 4x increase in range over Bluetooth 4.2
- Mesh-enabled networking (with an infinite theoretical range)

With all the advantages of mesh networking, Bluetooth 5 is now be able to communicate over longer distances. Not only will the packet be relayed over longer distances, but the broadcast messages should also be able to relay over longer distances. This means that even if the device is not close to you, you will be able to hear its broadcast message as the middle nodes would be relaying it.

 The Bluetooth mesh specification v0.9 document was released in December 2016, but it doesn't provide the description of routing/flooding protocols that BLE mesh uses. A complete Bluetooth core specification v5.0 was released in December 2016 and is available at `https://www.bluetoot h.com/specifications/adopted-specifications`.

Security complications in BLE mesh

Just like any other mesh network, a Bluetooth Low Energy mesh faces many security threats. Mesh networks are very useful for small and large applications. In mesh networks, each intermediary node acts as a router to transmit traffic among peers if the destination node is not directly connected to the sender host. The formation of these nodes is considered reliable, and because of this, it is very hard to detect the presence of an intruder in the network.

BLE mesh security is becoming a hot issue among researchers as it is open to many attacks. As Bluetooth Low Energy promises a strong future in IoT, it is becoming very popular. But without providing concrete prevention against threats, it will be very difficult for it to make its presence in the industry. A Bluetooth Low Energy mesh is open to a variety of attacks, such as **man-in-the-middle** (**MITM**), passive and active eavesdropping, and privacy and **denial of service** (**DoS**) attacks. Bluetooth v4.2 catered to many of these issues by providing a better security mechanism (encrypted link using the Diffie-Hellman key exchange algorithm). However, BLE meshes are still open to battery exhaustion attacks.

A battery exhaustion attack is an attack that consumes a device's battery to disable it. In the case of a Bluetooth Low Energy mesh, if there is an active battery exhaustion attack, it can take down a whole network. Since IoT devices are heavily dependent on batteries, this attack poses a serious threat on BLE meshes. It is a form of DoS attack through battery exhaustion.

Here are some forms of battery exhaustion attacks in BLE:

- **Service request power attacks**: The attackers bombard a BLE device by sending requests to particular GATT services to deplete the batteries
- **Malignant power attacks**: This creates or modifies an executable (to be run on the BLE device) to drain the battery and, in some cases, to run harmful Trojan horses on the target device
- **Benign power attacks**: This forces a device to repeatedly perform battery-heavy tasks to speed up battery drainage

CSRMesh

CSRMesh is a mesh protocol developed to run over Bluetooth Low Energy. It extends the traditional Bluetooth Low Energy functionality and enables it to relay messages and broadcast over longer distances. The protocol creates a mesh from CSR SoC-based (**CSR1010** and **CSR1011**) devices. This solution is very effective for a home automation system as it is not bound to a transmission range. CSR also markets its development kit, which helps in prototyping. Since it is a BLE-based device, you can connect through your Android, iOS, or desktop device. Here is the complete list of products offered by CSRMesh:

- Bluetooth smart starter development kit (`http://www.digikey.com/product-detail/en/DK-CSR1010-10169-1A/DK-CSR1010-10169-1A-ND/4514908`)
- Bluetooth smart remote control development kit (`http://www.digikey.com/product-detail/en/DK-CSR1011-10148-1A/DK-CSR1011-10148-1A-ND/4208929`)
- Bluetooth smart remote control evaluation kit (`http://www.digikey.com/products/en?KeyWords=DK-CSR1011-10147-1A%20%20&WT.z_header=search_go`)
- CSRmesh development kit (`http://www.digikey.com/products/en?KeyWords=DK-CSR1010-10184-1A&WT.z_header=search_go`)
- CSR1010 development kit (`http://www.digikey.com/products/en?x=21&y=13&lang=en&site=us&keywords=DK-CSR1010-10136-1A`)
- CSR1011 development kit (`http://www.digikey.com/products/en?KeyWords=DK-CSR1011-10138-1A&WT.z_header=search_go`)

- Bluetooth smart environmental sensor board (`http://www.digikey.com/produc ts/en?KeyWords=DK-ENV_SENS-10224-1A&WT.z_header=search_go`)

I recommend you buy a CSRMesh development kit that comes with at least three boards to make a full mesh structure.

 A complete list of devices with their distributors in each continent is available at `http://csrcorpkfxpa9lag2.devcloud.acquia-sites.com/p roducts/applications/sales-csr-solutions-provider-web-shops`.

Here is a view of a CSRMesh development board:

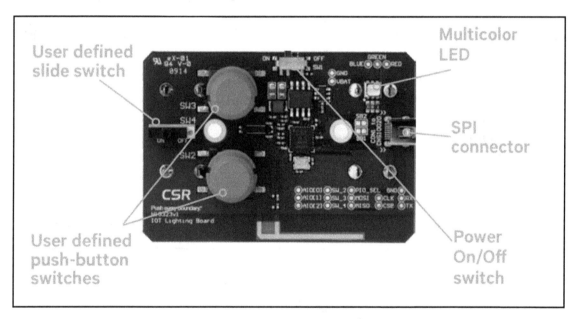

A CSRMesh development board comes with the following features:

- CSR1010 IC with EEPROM
- PCB antenna
- 1x RGB LED
- 2x buttons and 1x slide switch
- Power switch
- SPI programming connector
- 2x AA batteries

- I/O pads
- Temperature sensor

In 2016, Qualcomm acquired CSR, and it now sells CSR products while preserving the brand name. A list of CSR products can be found on the Qualcomm website, under the **Bluetooth Products** section
`https://www.qualcomm.com/products/bluetooth.`

Setting up the CSRMesh development kit

After unpacking your CSRMesh development kit, the next step is to know your board. The CSRMesh development kit comes with a uEnergy SDk to download the CSRMesh Light application on the development board. The CSRMesh development board interface consists of:

- **Switch SW1**: Power switch.
- **Button SW2**: Pressing this button for 2 seconds will remove the board from the mesh network. The behavior can be turned off by user configuration.
- **Button Sw3** and **SW4**: Unassigned.
- **RGB LED**: Blinks blue until it is not assigned, blinks yellow when association is in progress, blinks red when in fault, and the light color can be controlled by the user once the association is made.

A QR code representing a CSRMesh device tag can be found with the development board:

You can find the following information on the service tag:

- **Bluetooth Address** (**BT**) and crystal trim values (**XTAL TRIM**). Set these values in `light_csr101x_A05.keyr`, provided with the development kit, before programming.
- **SN** represents the serial number of the board.
- The **Universally Unique Identifier** (**UUID**) and **Authorization Code** (**AC**) are used by control applications during associations.
- The QR code is another representation of the UUID and AC.

 The URL to download CSR uEnergy tools is provided once you buy the board from the CSRMesh website.

Follow these steps to program service tag information (UUID, AC, and so on) in the **non-volatile memory** (**NVM**) of the board:

1. Open Command Prompt.
2. Program the UUID and AC value in the EEPROM using the serial interface link. If the base `NVM_START_ADDRESS` is defined in the `.keyr` file, use the following command to program the AC and UUID (use Little Endian-hex representation):

```
<CSR_uEnergy_Tools Path>uEnergyProdTest.exe -k light_csr101x_A05.keyr
-m1 0x04 <ENTER_UUID> <ENTER_AC>
```

3. If the `.keyr` file is not included in the command, the device UUID and authorization code can be programmed as follows:

```
<CSR_uEnergy_Tools Path>uEnergyProdTest.exe -m1 0xF804 <ENTER_UUID>
<ENTER_AC>
```

Here is a view of the non-volatile memory of CSRMesh:

Entity name	Type	Size of entity (words)	NVM offset (words)
Sanity word	`uint16`	1	0
Application NVM version	`uint16`	1	1
CSRMesh device UUID	`uint16 array`	8	2

CSRMesh device authorization code	uint16 array	4	10
CSRMesh network key	uint16 array	8	15
CSRMesh device ID	uint16	1	25
CSRMesh message sequence number	uint32	2	26
CSRMesh device eTag	uint16 array	4	28
CSRMesh association state	Boolean	1	32
Light model group IDs	uint16 array	4	33
Power model group IDs	uint16 array	4	37
Attention model group IDs	uint16 array	4	41
Data model group IDs	uint16 array	4	45
Light RGB and power values	structure	2	47
Bearer model data	structure	3	50

The Android CSRMesh control application

CSR offers a control application (on Android and iOS) that communicates with BLE mesh devices. The application connects to the mesh device that runs CSR's custom-defined CSRMesh control profile. The control application can be used to:

- Set up a mesh network by associating individual mesh devices
- Configure and cluster the mesh devices in a mesh network
- Control light by setting colors and brightness levels

Download the Android application provided with this chapter and run in debug mode on your Android device. An alternate method is to copy the CSRmeshDemo.apk file and install it manually. Make sure to check **Unknown sources** in the settings of your Android phone.

Step 1 - connect to the network

Once you start the app after installation, you will get a **Connecting...** dialog box on your screen:

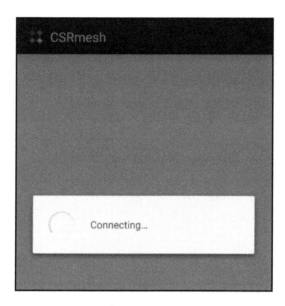

Step 2 - set up a mesh network

In order to set up a mesh network, the application needs to generate a passphrase. Follow these steps to generate a passphrase and set up the network:

1. Go to the **Security Settings** page, enter a passphrase, and press **OK**. This is an authorization key for your network and will be used by new devices to connect to your mesh network.
2. The **Security Settings** screen will now show the **Bluetooth Address** (**BA**) of the connected bridge device.

It is recommended you use the authorization code for every associating device. However, you can find a checkbox that disables the device authorization feature. This feature should only be used if you are using an old version of the kit and the devices do not support the AC feature.

Step 3 - associate new devices with your network

Choose **Device Association** from the drop-down menu, and it will open a screen with a list of available devices for association. This list will also show the 128-bit UUID of the devices:

Long-press on the device you want to associate with the network. A message in the bottom will show a progress bar with the association process. Once it is done processing, it will show a toast message that the device has been added to your network. You can repeat the process with every device you want to connect to your network.

Step 4 - authorize connected devices

At this point, a QR code will come in handy as it is required to make the associations. At the bottom of the **Device Association** screen, you can see a **QR Code** button. Press it and scan the QR code using your device camera.

By this point, you have successfully associated and authorized the devices on your mesh network. The CSRMesh application also provides the functionality to control devices through this app.

Additional step - controlling light and thermostat

Navigate to the **Light Control** or **Temperature Control** screens to get a complete list of control options. You can choose the color using a color wheel and associate that color to an individual or group of lights. Similarly, you can control the thermostat by tapping up and down on the screen. This control can be run over an individual or a group of devices.

Here is a view of the control screens:

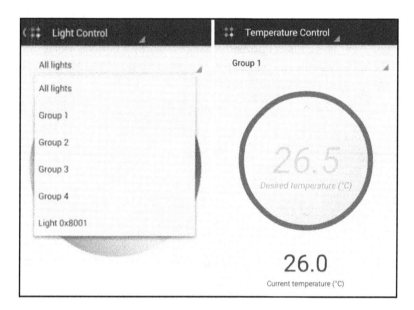

Additional step - configuring your devices

Navigate to **Configuration** from the drop-down menu. By default, the app creates four groups. You can get additional options by long-pressing on any group. Renaming, deleting, device information, firmware information, and request data options will be visible in the pop-up menu.

Additional step - grouping your devices

In the **Configuration** menu from the dropdown, select the group you want to edit. It will give you all the available devices that you can associate with this group. You can check and uncheck the devices from the group and click on the **Apply** button at the bottom of the screen.

CSRMesh library for Android

This section describes the CSRMesh library for Android so that you can learn to make your own control apps. We will take a look at the functionalities offered by this library and how it can be used in an Android application. You can also refer to the CSRMesh demo source code provided with this chapter.

The CSRMesh library for Android comes with `MeshService` and Model classes dedicated for different tasks. This `MeshService` represents an Android service that provides methods to connect to a bridge and its associated devices. It can be considered as the core of the CSRMesh Android API, which is responsible for performing central-peripheral communication. Once you make a connection, the next step is to send control commands to the mesh network. This can be performed using Model classes.

In order to understand the different components of the CSRMesh Android library, refer to this block diagram:

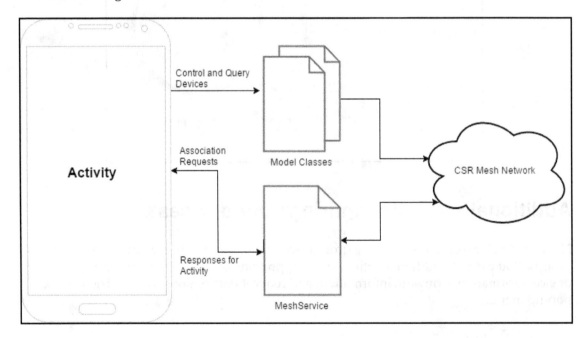

Getting started with the CSRMesh Android library

The CSRMesh library is not yet available on Gradle, so you cannot add it directly to the `build.gradle` file in your Android project. You need to add the `CSRmeshLibrary.aar` file in your project. You also need to add **Spongy Castle** libraries used for mesh security. These files can be found under `CSRMeshAndroidApplicationCSRmeshDemoCSRmeshLibrary`, provided with this chapter.

After attaching the library, go to the `AndroidManifest.xml` class to add a `MeshService`. This can be achieved by writing this in your `Manifest.xml` file under the `<application>` tag:

```
<service
    android:name="com.csr.mesh.MeshService"
    android:enabled="true"
    android:exported="false" >
</service>
```

Also, provide the `BLUETOOTH` and `BLUETOOTH_ADMIN` permissions in your application using the following lines of code:

```
<uses-permission android:name="android.permission.BLUETOOTH" />
<uses-permission android:name="android.permission.BLUETOOTH_ADMIN" />
```

Using MeshService

Since `MeshService` is the core class of the CSRMesh Android library, you will need it in your main activity. Create a global variable and bind it inside the `onCreate` method:

```
private MeshService mMeshService;

@Override
protected void onCreate(Bundle savedInstanceState) {
super.onCreate(savedInstanceState);
// Make a connection to MeshService to enable us to use its services.
    Intent bindIntent = new Intent(this, MeshService.class);
    bindService(bindIntent, mServiceConnection, Context.BIND_AUTO_CREATE);
}
```

If you notice, there is an unknown variable in the previous code, `mServiceConnection`. This is an object that will handle the service connection. Declare it using the following code:

```
private ServiceConnection mServiceConnection = new ServiceConnection() {
public void onServiceConnected(ComponentName className, IBinder rawBinder)
{
mService = ((MeshService.LocalBinder) rawBinder).getService();
    }

public void onServiceDisconnected(ComponentName classname) {
mService = null;
    }
};
```

`MeshService` is now ready, but the CSRMesh stack has not yet been initialized. We will first declare a handler that is responsible for receiving messages from CSRMesh (which can later be used in the user interface of the application). In order to do this, set a `mMeshHandler` in `mMeshService`:

```
private final Handler mMeshHandler = new Handler(this);

private boolean connect(){
    mMeshService.setHandler(mMeshHandler);
}
```

The next step is to start a Bluetooth scan to detect CSRMesh devices in the vicinity. For this, we need to declare a `ScanCallBack` object, which will be called as CSRMesh library hears the scan response:

```
private LeScanCallback mScanCallBack = new LeScanCallback() {
@Override
    public void onLeScan(BluetoothDevice device, int rssi, byte[]
scanRecord) {
        mMeshService.processMeshAdvert(device, scanRecord, rssi);
  }
};

private boolean connect(){
 mMeshService.setHandler(mMeshHandler);
 mMeshService.setLeScanCallback(mScanCallBack);
}
```

This `mScanCallback` function must explicitly call `mMeshService.processMeshAdvert()` for every advertisement it receives. This will allow the library to consume any CSRMesh advertisement packets. The library is smart enough to automatically enable and disable Bluetooth scanning according to your requirements. But if continuous scanning is required, you can do it by calling this method:

```
mMeshService.setContinuousLeScanEnabled(true);
```

By default, BLE notifications are enabled on the bridge, but you can manually enable it by calling this method:

```
setBridgeNotificationsEnabled(true);
```

After scanning, you can connect the library to the bridge by calling the following:

```
mMeshService.connectBridge(bridgeDevice);
```

The `bridgeDevice` variable is the `BluetoothDevice` object that represents a bridge device. Scanning for a bridge device is the same as scanning for BLE devices. Since you want to filter only CSRMesh devices in your scan, you can use the bridge service UUID, `0xFEF1`.

Mesh handler

If you look at the previous code, we associated a handler object with `mMeshService` that is responsible for receiving responses from CSRMesh. It is better to extend a custom class that inherits the handler class for your custom implementation. The messages received by the handler are defined in the `MeshService` class. For example, when an association message is caught by the handler, it will have the value of `MeshService.MESSAGE_DEVICE_ASSOCIATED`. So in order to write a handler for this kind of response, you need to write code like the following:

```
public void handleMessage(Message msg) {
switch (msg.what) {
case MeshService.MESSAGE_DEVICE_ASSOCIATED :
            Do something here...
            break;
    }
}
```

Since these are messages coming back from CSRMesh devices, they might have some data associated with them. In order to retrieve that data, you can use tags like the following:

```
case MeshService.MESSAGE_DEVICE_ASSOCIATED: {
// New device has been associated and is telling us its device id.
    // Request supported models before adding to DeviceStore, and the UI.
    int deviceId = msg.getData().getInt(MeshService.EXTRA_DEVICE_ID);
int uuidHash = msg.getData().getInt(MeshService.EXTRA_UUIDHASH_31);
    Log.d(TAG, "New device associated with id " + String.format("0x%x",
deviceId));
break;
}
```

Mesh controlling using Model classes

The application can send control requests using Model classes. The CSRMesh library provides various Model classes, and they can be acquired by calling the appropriate method on the `MeshService` class. Here is the list of Model classes available in the CSRMesh library:

- `ActuatorModelApi`: Send requests to the device that are not awake 100% of the time
- `AttentionModelApi`: Tell a device to get user attention
- `BearerModelApi`: Enable or disable the relay functionality
- `ConfigModelApi`: Acquire device configuration and remove associations
- `DataModelApi`: Exchange manufacturer's data using ACK and UNACK packets
- `LightModelApi`: Control and modify lighting information
- `GroupModelApi`: Manage a mesh group
- `PowerModelApi`: Manipulate or acquire a mesh device's power state
- `FirmwareModelApi`: Perform firmware-related operations
- `PingModelApi`: Ping the mesh network just like a network ping
- `SensorModelApi`: Control and modify sensor information

You can use these classes in order to perform tasks on mesh devices. For example, if you want to get information about a device, you can use this:

```
ConfigModelApi.getInfo(mDeviceId,
ConfigModelApi.DeviceInfo.VID_PID_VERSION);
```

Here, `mDeviceId` is the ID of the device you need to acquire information from. Similarly, if you want to obtain the firmware version of a particular device, you can write this:

```
FirmwareModelApi.getVersionInfo(mDeviceId);
```

An example of controlling a mesh device is to set the color of the light. You can call `LightModelApi` to set RGB colors of the light:

```
LightModelApi.setRgb(mDeviceId, red, green, blue, (byte)0xFF, 1, false);
```

The `(byte)0xFF` variable identifies the light level, `1` identifies the duration, and `false` describes whether the command is to be acknowledged or not.

By going through each Model class, you will be able to perform a range of functionalities over mesh devices. The goal of this chapter is to provide enough understanding of the API so that you can implement your own projects. I encourage you to explore the Model classes and implement a range of controls.

Summary

This chapter dealt with the concept of mesh networks and how Bluetooth mesh networks are beneficial over conventional Bluetooth Low Energy. ZigBee was discussed as a competitor of Bluetooth Low Energy in mesh technology. We then had a brief introduction to security complications in Bluetooth mesh networks. The goal of this chapter was to provide you with a practical approach, so you can implement your own project. Therefore, the CSRMesh API was discussed in detail, where we saw a step-by-step approach to implementing Android control for a CSRMesh network.

In the next chapter, we will learn another practical implementation of BLE, an Internet-Bluetooth gateway using the Raspberry Pi 3.

7
Implementing a Bluetooth Gateway Using the Raspberry Pi 3

We are on the verge of a technological boom, where the concept of smart homes will lead the way. If you analyze the history computer technology, you can define it in three periods: the era of personal computing, the era of cellular phone evolution, and the era of smart devices. While this chapter does not describe the ups and downs of each era, it is important to know that there was no gap between these periods. They came back to back, and companies that did not catch up were left behind.

A notable example from the first period is IBM: a company that stood in competition with Apple in the early era of computing now does not list itself among the major brands of personal computers. Nokia, a well-known name in cellular phone manufacturing, could not keep the pace with Apple and Samsung in the second era. It stuck with the Symbian OS when Apple was developing the state-of-the-art iOS and Samsung was working with Android. Nokia decided to go for the Microsoft Windows OS for their smartphones but could not reclaim its former glory. The company was later acquired by Microsoft and operated under Microsoft Mobile.

The history of Nokia from the second era is unfortunate, but in October 2014, Nokia announced its intentions to re-enter the consumer electronics market, and in January 2017, they released their first Android-powered cellphone, **Nokia 6**, at the Mobile World Congress 2017.

We are living in the third period right now and entrepreneurs know it. Although we are still waiting for the Apple of third era, some companies are doing an amazing job in bringing the technology upto speed. It is a given that the companies who understand the potential of smart devices will win in the next decade. Since technology is available at low cost, companies are manufacturing small computers that are low powered and extremely powerful for the price. This chapter will have a special focus on the Raspberry Pi, made by the Raspberry Pi Foundation. In this chapter, we will go over:

- Introducing the Raspberry Pi 3 and its potential use in IoT applications
- Setting up a Linux operating system on the Raspberry Pi 3
- Implementing a gateway application using RESTFul APIs on Node.js
- Implementing a simple website to access Bluetooth devices on the Web

Introducing the Raspberry Pi

The Raspberry Pi is a smart microcomputer developed by the Raspberry Pi Foundation, a UK-based company. It was the first company to successfully provide cheap ARM-based computers for educational purposes, focusing on educational use and use in developing countries. The intent of this company is to promote the teaching of computer science in schools. The company struggled in its early days, when the prices of their computers were high, but the founders--Eben Upton, Rob Mullins, Jack Lang, and Alan Mycroft--collaborated with Pete Loman, managing director of Norcott Technologies, to form the Raspberry Pi Foundation.

As of 2016, the Raspberry Pi Foundation has successfully sold over 10 million Raspberry Pis, making it the bestselling British computer to date. The Raspberry Pi has gone through several evolutions as well. The evolution of the Raspberry Pi can be seen here:

- Raspberry Pi 1 Model B
- Raspberry Pi 1 Model B+
- Raspberry Pi Zero
- Raspberry Pi 2
- Raspberry Pi 3 Model B
- Raspberry Pi Zero W

They all feature a Broadcom **system-on-chip** (**SoC**), which includes an ARM-based CPU and an on-chip GPU. A view of the Raspberry Pi family is shown in the following image:

Raspberry Pi 3 Model B

In this chapter, we will use the Raspberry Pi 3 Model 3. It is the most powerful Raspberry Pi and belongs to the third generation of devices in its family. It replaces the Raspberry Pi 2 Model B and includes the following:

- 1.2 GHz 64-bit quad-core ARM processor
- 1 GB of RAM
- Four USB ports
- 40 GPIO ports
- 802.11n wireless LAN
- Bluetooth 4.1

- Full HDMI and Ethernet ports
- MicroSD card slot
- 3.5 mm audio jack
- Camera and display interface
- VideoCore IV SoC GPU core

Unlike the Raspberry Pi 2, the Pi 3 Model B has essential wireless technologies directly on the board. Wi-Fi and Bluetooth Low Energy are featured on the board, which makes it an ideal standalone computer without the need for any additional dongles. The Raspberry Pi Foundation suggests you use the Pi 3 Model B for standalone projects and Pi Zero W for embedded use.

The Raspberry Pi 3 Model B can be configured with various operating systems, but Raspberry Pi encourages the use of Raspbian, a Debian-based Linux operating system. Other operating systems compatible with the computer are:

- Ubuntu Mate and Ubuntu Core (Snappy)
- Windows 10 IoT Core
- RISC OS
- Fedora
- Kali Linux
- Slackware
- Android Things

In this chapter, we will go with the Raspberry Pi Foundation's recommendation and use Raspbian as our primary operating system.

Applications of the Raspberry Pi in the Internet of Things

The Raspberry Pi 3 is a powerful computer for small-scale applications. It features a 1.2 GHz quad-core processor with 1 GB of RAM, good enough spec for low-power applications. There are many smart-home devices that you can buy, but making your own IoT device is always exciting. The Pi 3 is the perfect candidate for this purpose. In this section, we will discuss the potential IoT projects you can implement using the Raspberry Pi.

Media center using Raspberry Pi

Creating something that can act as your very own media center is always exciting. The $35 Pi 3 is perfect for building your own media center, which can act as your streaming device (for video at 720p and audio). It has its limitations as you can't use it for 1080p HD, but for a cheap price, it offers good enough resolution for your media. But it can also stream music from radio, Spotify, Pandora, Google Music, and other streaming platforms.

Cloud storage using Raspberry Pi

Google Drive, Microsoft OneDrive, Dropbox, and other services provide cloud storage, but usually they are not free. You can build your own cloud storage with a Raspberry Pi and an external SSD. Since the Model 3 comes with built-in wireless LAN and Ethernet capabilities, you can make your own cloud storage service that you can access from anywhere.

Tracker using Raspberry Pi

Another interesting idea is to make your own tracker. It requires a separate GPS module that you can, say, put on your pet and let them roam around freely. The tracker will keep you alerted about the location of your pet. For this project, you will need your own battery as well.

Web server using Raspberry Pi

If you are a developer and want to build a web server, you can use the Raspberry Pi. An Apache web server instance on a Linux-based Raspberry Pi is ideal for web developers. You can also connect a temperature/humidity sensor to keep track of your server's conditions.

Gateway for Bluetooth devices using Raspberry Pi

In this chapter, we will dive deep into using the Raspberry Pi as a Bluetooth gateway. Imagine a device that is constantly monitoring the Bluetooth IoT devices in your home and reporting to you. The Raspberry Pi 3 can act as a gateway that monitors Bluetooth devices and informs you about them. Imagine a scenario where you have an Estimote Bluetooth beacon with motion sensors and you put it on your front door. If someone breaks in, the beacon will sense it, and the Raspberry Pi will broadcast a notification using the Internet.

When you merge two technologies (Wi-Fi and Bluetooth), the possibilities and the implementations are limitless. Luckily, the Raspberry Pi 3 features both of these technologies on a single board, with a 1.2 GHz quad-core processor running to support them. The goal of this chapter is to give you a strong hold on the basics of Bluetooth Gateway. With the help of Chapter 2, BLE Hardware, Software, and Debugging Tools, Chapter 3, *Building a BLE Central and Peripheral Communication System*, and this chapter, you will be able to implement many real life applications on your Raspberry Pi 3. In later sections, we will see a step-by-step guide to set up your Raspberry Pi as a gateway to listen to Bluetooth Low Energy devices.

Set up your Raspberry Pi with Raspbian

In order to start setting up your Raspberry Pi, you need to first install the Raspbian OS. Raspbian is a Debian-based Linux operating system for the Raspberry Pi. It was created in June 2012 by Mike Thompson and Peter Green as an independent project, but now the Raspberry Pi Foundation recommends it. This book will teach you how to install Raspbian using **New Out Of the Box Software (NOOBS).** NOOBS is an operating system installer that contains Raspbian. In addition to this, it also provides an alternative OS in case you don't want to use Raspbian.

 You can download NOOBS from https://www.raspberrypi.org/downlo ads/noobs/.

After downloading NOOBS, extract it. Select all the files in the extracted directory and copy them to an SD card. The Raspberry Pi 3 comes with a built-in microSD card slot. You can connect your microSD card into an SD card jacket and copy the files after connecting it to your computer. Once they are copied, insert the card in the Raspberry Pi 3 and power up the Pi. Once you have successfully powered it on, you will see the Raspberry Pi startup screen. Install Raspbian from this screen:

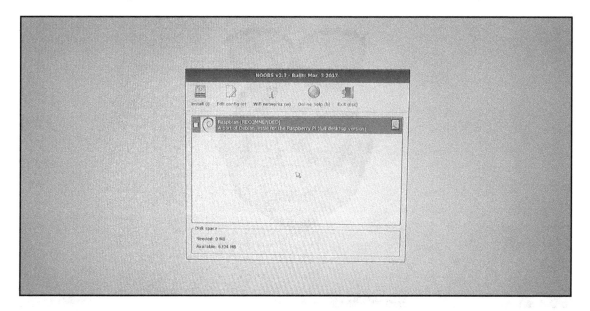

It will take approximately half an hour to fully install the operating system. Before that, make sure that you have:

1. HDMI Connection with a monitor.
2. Keyboard and mouse connection using USB.
3. Power Connection using a 5V power adapter.

Once the Raspbian operating system is running on your Raspberry Pi, it should show the following screen:

Your Raspberry Pi is now all set up and ready to be used.

 You can SSH into your Raspberry Pi once it is connected. This book will assume that you are directly working on your Raspberry Pi and have attached a keyboard and a mouse to it. Make sure that you have Internet connectivity before moving to the next section.

Deploying a Bluetooth gateway on your Raspberry Pi

The goal of this project is to convert your Raspberry Pi into a Bluetooth gateway so that it can scan and inform you about the surrounding Bluetooth devices (including their services and characteristics). The prerequisites of this project are:

- A Raspberry Pi 3 Model B (or Raspberry Pi 2 with Wi-Fi and Bluetooth dongles)
- An Internet connection

Let's start by setting up the required packages on the Raspberry Pi. I will assume that you are a Linux novice and will explain every step in detail.

Step 1 - Update and upgrade the Pi

The first step before starting development on your Raspberry is to make sure that it is up to date. To do this, run the following command:

```
sudo apt-get update
```

This command will update the package lists from the repositories and update them to get information on the newest versions of packages and their dependencies. The next step is to upgrade the installed packages to the latest packages. To do this, run the following command:

```
sudo apt-get upgrade
```

Both of these commands will be run from the Linux terminal. This can be found on the top bar in your desktop.

Step 2 - Installing Bluetooth and Node.js

The next step is to install the essential Bluetooth and Node.js packages. For Bluetooth communication, you will require three packages:

- Bluetooth
- BlueZ
- libbluetooth-dev

These packages are essential as the Bluetooth gateway will use them to communicate with Bluetooth devices. These steps are also necessary if you are working with an external Bluetooth USB dongle.

 You can find further details about BlueZ at `http://www.bluez.org/`.

The next step is to install Node.js on your Raspberry Pi. Since the gateway application is written in JavaScript, you will need Node.js to run the application. The easiest way to install it on a Debian-based Linux device is with this set of commands:

```
curl -sL https://deb.nodesource.com/setup_6.x | sudo -E bash -
sudo apt-get install -y nodejs
```

The `curl` command normally comes with Raspbian, but if it is not there, you can install it with this command:

```
sudo apt-get update && sudo apt-get install php5-curl
```

To confirm that node is successfully installed and running, you can check the Node version by typing this:

```
node -v
```

In my case, the version check gave me `v6.10.0`. The last step in setting this up is to install `node-gyp`. You can do so by running the following command:

```
sudo npm install -g node-gyp
```

Restart the Raspberry Pi:

```
sudo shutdown -r now
```

Step 3 - Getting the project folder

In order to get the whole project together, refer to the book's code files and find the folder called `Bluetooth Gateway`. Copy the folder somewhere on your Raspberry Pi Desktop using a standard USB. Assuming that your downloaded files now rest in the `~/Desktop/BluetoothGateway` folder, you should be seeing two folders there. One is `gateway` and the other is `navible`. Before moving into the technicalities, make sure that you have all the project dependencies installed. Navigate to your `gateway` folder and run the following command:

```
cd gateway
sudo npm install
```

This will install all project-related dependencies and keep them in the `node_modules` folder. In order to install the server, use the following command:

```
sudo npm start
```

Upon starting the project, you should see the following screen:

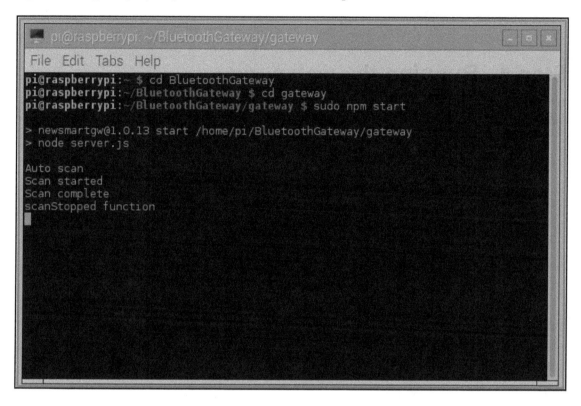

If you are unable to see this screen, there might be a problem in communicating with the Bluetooth module. To troubleshoot this problem, install `bluez-hcidump` by executing the following command:

```
sudo apt-get install bluez-hcidump
```

Once it has been installed, check whether the `hci0` interface is up and running. To do this, run the following command:

```
sudo hcidump
```

If it is unable to find the interface, you can manually restart it by executing the following:

```
sudo hciconfig hci0 down
sudo hciconfig hci0 up
sudo bluetoothctl
```

You should now start the discovery process by running this command:

```
bluetooth# scan on
```

Step 4 - Running the web server

We are now ready to run the servers. Go to the terminal and navigate to the project folder called `/BluetoothGateway/navible` and get the dependencies:

```
sudo npm install
sudo npm start
```

This should give you the URL where the web server is running. In our case, the server is running on `https://192.168.1.111:8000`. It will ask for the username and password to access it. The default username and password are both `user1`. After authenticating, you should be able to scan the device by pressing the **Scan** button:

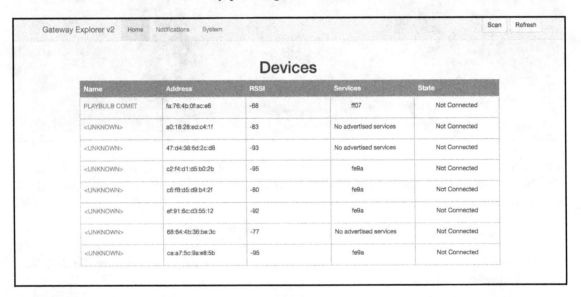

This shows the list of all Bluetooth devices available in the vicinity. In the next section, we will study the deployed Bluetooth application in detail.

Implementing a Bluetooth gateway on the Raspberry Pi

In this section, we will study the Bluetooth gateway server (as described in `~/Desktop/BluetoothGateway/gateway`). We will study the RESTful services offered by the gateway and we will jump into the gateway code. The gateway server, as described in the previous section, is a web server that runs on Node.js and implements GAP, GATT, and RESTful web services. In order to get a recap of the Bluetooth GAP and GATT architecture, you can refer to `Chapter 1`, *BLE and the Internet of Things*.

 The code for the gateway can be found in the `gateway` folder of the project source. The core API used by this project is developed by the Bluetooth SIG, and the detailed documentation, including whitepapers, can be found at `https://www.bluetooth.com/develop-with-bluetooth/white-papers`.

The architecture of a Bluetooth gateway consists of three main components:

- Bluetooth nodes (or smart devices)
- Bluetooth gateway/server (Raspberry Pi)
- Internet-enabled client (such as a cellphone or web browser)

The functionality of these three components can be understood by considering an example smart home. You have a temperature sensor in a Bluetooth beacon that is constantly broadcasting the temperature information about a room to its surroundings. In this case, the Bluetooth beacon is one of the Bluetooth nodes. A Raspberry Pi acting as a Bluetooth gateway/server is listening to all the devices in its surrounding and making that information available to the web server. You are somewhere outside your house but have a customized application installed that polls the server to obtain the information once every 5 minutes. In this case, your cellular phone is an Internet-enabled client that is listening to the server for potential informational messages. Imagine that an abrupt or unusual change in temperature is detected by the Bluetooth beacon. The information is then passed to the gateway, which will make this information available to potential clients. If a client is listening to the server, it will generate an alert, notifying the user that there is a potential fire in the apartment. This is how a complete Bluetooth gateway cycle is performed from a local Bluetooth network to a notification message on a client's cellular phone.

Here is a visual representation of the Bluetooth gateway architecture:

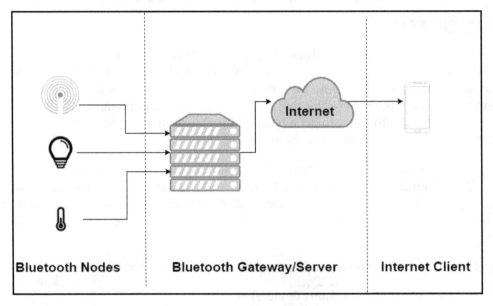

Bluetooth SIG GATT server API

The Bluetooth SIG defines a RESTful GATT server API for allowing remote Internet clients access to Bluetooth nodes through a Bluetooth gateway. An architectural diagram provided by the SIG emphasizes the terminologies:

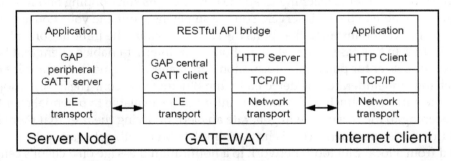

Server Node refers to our architectural distribution of Bluetooth nodes, **GATEWAY** represents the Bluetooth gateway/server block, and **Application** represents an **Internet client** listening to the gateway. In this section, we will focus on the services offered by the Bluetooth gateway. These services are then used by the Bluetooth HTTP server to perform various operations, such as Bluetooth scans, service discovery and characteristics, and reading/writing.

RESTful services offered by the RESTful smart server API

The RESTful smart server API exposes many RESTful services that help the client communicate with Bluetooth nodes. These services include the following:

REST action	URL	Description
GET	`/gap/nodes`	Get information about the Bluetooth devices in the proximity
GET	`/gatt/nodes/{nodesId}/service`	Discover services related to a Bluetooth device
GET	`/gatt/{userId}/nodes/{nodesId}/services/{servicesId}/characteristics`	Get a list of all the characteristics related to a service
GET	`/gatt/{userId}/nodes/{nodesId}/services/{servicesId}/characteristics/{characteristicsId}/value`	Get a value of a characteristics
PUT	`/gatt/{userId}/nodes/{nodesId}/services/{servicesId}/characteristics/{characteristicsId}/value`	Write a value to a characteristics
GET	`/management/nodes`	List of devices with their pairing status
GET	`/management/nodes/{nodeId}`	Get a device's pairing capabilities
GET	`/management/forcescan`	Initiate a force scan from the gateway
GET	`/management/{userId}/clearbondedstore`	Clear the bond and store the currently bonded devices
PUT	`/management/nodes/{nodeId}/getInfo`	Get information about a Bluetooth device
PUT	`/management/{userId}/nodes/{nodeId}/pair`	Pair with a Bluetooth device
GET	`/management/{userId}/nodes/{nodeId}/unpair`	Unpair with a Bluetooth device

GET	`/management/{userId}/nodes/{nodeId}/unbond`	Unbond with a Bluetooth device
GET	`/management/{userId}/{nodeId}/passKey/value/{value}`	Transfer a passkey to the Bluetooth device
GET	`/management/{userId}/{nodeId}/pair-input/{confirmed}`	Set the pairing confirmed status to the Bluetooth device

For these RESTful services, `userId` is the unique identifier attached to a user's gateway login. This is part of the authentication mechanism the Bluetooth SIG provided for validating incoming calls. The SIG had to extend a Node.js Bluetooth Low Energy library called `noble`. They modified the library in order to implement the pairing protocol. All the modifications can be found in the `our_module` folder in the `gateway` project folder.

Important libraries and their usage

In the gateway project, a majority of the communication is done by a Node.js library named `noble`. This library is responsible for providing any Bluetooth-related tasks (other than pairing, as it is done with the Bluetooth SIG's implementation). The main reference for this can be found in the `README.md` file:

```
var noble = require('noble');
```

If you read further along in the `our_modules/README.md` file, you can find various Bluetooth calls associated with the `noble` variable. A short list of noble commands defined in the `our_modules/README.md` file is as follows:

```
javascript
noble.startScanning(); // any service UUID, no duplicates


noble.startScanning([], true); // any service UUID, allow duplicates


var serviceUUIDs = ["<service UUID 1>", ...]; // default: [] => all
var allowDuplicates = <false|true>; // default: false

noble.startScanning(serviceUUIDs, allowDuplicates[, callback(error)]); //
particular UUID's
```

```
__NOTE:__ ```noble.state``` must be ```poweredOn``` before scanning is
started. ```noble.on('stateChange', callback(state));``` can be used
register for state change events.
```

```
#### Stop scanning

```javascript
noble.stopScanning();
```

Peripheral

Connect

```javascript
peripheral.connect([callback(error)]);
```

Disconnect or cancel pending connection

```javascript
peripheral.disconnect([callback(error)]);
```

Update RSSI

```javascript
peripheral.updateRssi([callback(error, rssi)]);
```

Discover services

```javascript
peripheral.discoverServices(); // any service UUID

var serviceUUIDs = ["<service UUID 1>", ...];
peripheral.discoverServices(serviceUUIDs[, callback(error, services)]); //
particular UUID's
```

For a complete list of commands, you can refer to the `our_modules/README.md` file
provided in the `gateway` folder of the project source. These commands are then used in
different parts of the `gateway` project in order to communicate with the Bluetooth nodes.
Along with the `noble` calls, you can also find the structure of a `peripheral` object. It is
defined as follows:

```
javascript
peripheral = {
  id: "<id>", //unique address of the Bluetooth Device
  address: "<BT address">, // Bluetooth Address of device, or 'unknown' if
not known
```
```

```
 addressType: "<BT address type>", // Bluetooth Address type (public,
random), or 'unknown' if not known
 connectable: <connectable>, // true or false, or undefined if not known
 advertisement: { //advertisement packet
 localName: "<name>", //name of the device advertising the packet
 txPowerLevel: <int>, //transmission power level
 serviceUuids: ["<service UUID>", ...], //Unique Identifiers of all the
services offered by the peripheral
 manufacturerData: <Buffer>, //Manufacturer Data is the customized data
provided by the vendor of the peripheral
 serviceData: [
 {
 uuid: "<service UUID>" //UUID of a service
 data: <Buffer> //data of a service
 },
 ...
 ]
 },
 rssi: <rssi> //RSSI strength
};
```

Important JavaScript in the gateway project

In this section, we will quickly take a look at the important classes responsible for communicating with Bluetooth devices using the `noble` library. The BLE adapter handles all major BLE connections and traffic. This means that most classes call this if Bluetooth communication is required. You will also find references to the `noble` API and the customized Bluetooth SIG implementation of `our_modules` in this class. The main imports in this class are as follows:

```
var keystore = require('./keyStore');
var debug = require('debug')('bleAdapter');
//var error = debug('gateway:ble:error');
//var log = debug('gateway:ble:log');
var util = require('util');
var db = require('./dataModel');
var conts = require('./constants');
var timeconts = require('./timeconstants');
var events = require('events');
var Peripheral = require('./our_modules/noble/lib/peripheral');
var noble = require('./our_modules/noble');
var appserver = require('./app');
var bleNotify = require('./bleNotification');
var fs = require("fs");
```

The main module defines the major callbacks: which function to trigger if the device state changes, which function to call when scanning starts, and so on. You can customize these methods if you want different behavior to what is already written. Here is a brief guide on the functions written in this module:

- onScanStarted: A callback upon starting a Bluetooth scan
- onScanStopped: A callback upon stopping a Bluetooth scan
- onDiscoveredDevices: A callback after finding a Bluetooth device
- onNotification: A callback on receiving a notification from a characteristic
- onDeviceDisconnect: A callback on disconnecting a Bluetooth device
- connectDevice: A function that connects to a Bluetooth device

For a complete list of methods offered by this class, refer to the bleAdapter.js class:

- The dataModel.js file contains functions to form data objects that will go back from the server. These data objects can be defined in nodes, services, characteristics, managers, NodesJson (complete JSON of a Bluetooth device), ServicesJson (complete JSON of a service), characteristicsJson (complete JSON of a characteristic), and ManagementJson, which will incorporate all other JSONs to form a big payload.
- The management.js file provides all the functionalities related to the management of a Bluetooth device. If you refer to the RESTful table as mentioned previously, all the calls starting with /management/... are handled in this .js file. This class primarily depends on the bleAdapter.js (for any communication with the Bluetooth devices) and dataModel.js (to prepare objects to return back).
- The services.js file contains all the GATT-related RESTful calls. For example, the implementation of retrieving the services from a particular Bluetooth device can be found here. This class primarily depends on bleAdapter.js (for Bluetooth communication) and dataModel.js (preparing data objects to return back).

Implementing the Bluetooth gateway explorer v2

Until now, we've looked at the Bluetooth gateway/server implementation written in Node.js. In this section, we will study the gateway explorer v2 implementation of Bluetooth SIG. You can find the code in `~/BluetoothGateway/navible`. A gateway explorer is a simple web application written in Node.js. Its true potential lies in developing your own client application and then calling the gateway URLs but, for the sake of simplicity, we will define the standard Bluetooth client made by the SIG. The interface of the gateway explorer looks like this:

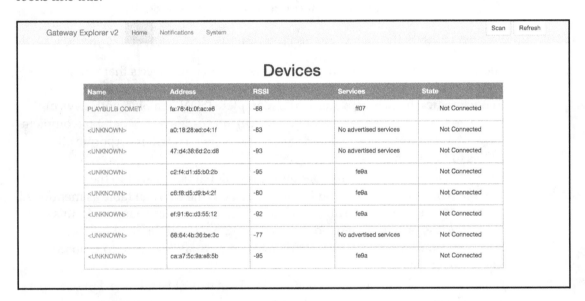

It is a single-page application with three tabs on the top and two buttons to provide scanning and refreshing functionalities. I'll assume you have web development knowledge and will only focus on the Bluetooth connectivity. In case you do not have web development knowledge, you can develop a client on any platform (for example, Android and iOS) and link it to the gateway IP address. For this application, the web server and Bluetooth gateway server are both running on the same Raspberry Pi, and that is why no URLs were used in the implementation.

Important libraries used in the web application

In the `navible` project, you will find a `package.json` file that lists all the dependencies related to this project. The project contains a complete implementation of a web-based client, including authentication. This is why you have to enter `user1` as both the username and password at the start of the application. The authentication in this project is achieved by the `hapi-auth-basic` library, which will be downloaded when you run `sudo npm install`. To host a web server, the project is using hapi version 13.2.2.

You will find other debugging and development utilities that are not significant from a Bluetooth communication point of view. The main point of communication for this web application is the `gateway` server running on the same Raspberry Pi.

Important JavaScript used in the web application

The `navible` project uses Node.js for developing the web application. It then uses jQuery to trigger web calls to the `gateway` server. For the sake of simplicity, we will ignore the web development done in the project. Since the main purpose of this book is to provide insight about Bluetooth communication, I will assume you have enough skills in web development to understand the HTML-JavaScript-CSS architecture.

Probably the most important folder in the `navible` project is `~/BluetoothGateway/navible/routes/`. It contains all the `BluetoothGateway` calls and their handlers. For example, all the calls for obtaining information about Bluetooth devices can be found in the `nodes.js` file, which exports the module as follows:

```
module.exports = [
  { method: 'GET', path: '/nodes', config: { auth: 'simple', handler:
nodes_handler }},
  { method: 'GET', path: '/nodes/{node_id}', config: { auth: 'simple',
handler: node_handler } },
  { method: 'GET', path: '/nodes/{node_id}/state/{connection}', config: {
auth: 'simple', handler: node_connection_handler } },
  { method: 'GET', path: '/nodes/closeallconnections', config: { auth:
'simple', handler: node_connection_closeall_handler } },
]
```

You can also find references to the Bluetooth gateway in the `nodes.js` class, providing strengths to the handler classes. For example, in the `node_handler` function, you can find an options call calling the `/gatt/nodes/{UUID}/services` GET request from the Bluetooth gateway:

```
var options = {
  path: '/gatt/nodes/' + node_uuid + '/services',
  method: 'GET',
};
```

Similarly, the Bluetooth communication related to GATT services can be found in the `services.js` file. The module contains two major gateway calls, `/nodes/{nodeId}/services/{serviceId}` and `/nodes/{nodeId}/clearservice/{serviceId}`, and exports as follows:

```
module.exports = [
  { method: 'GET', path: '/nodes/{node_id}/services/{service_id}', config: {
auth: 'simple', handler: service_handler }},
  { method: 'GET', path: '/nodes/{node_id}/clearservice/{service_id}',
config: { auth: 'simple', handler: clearservicenotifications_handler }}
  ]
```

The `characteristics.js` file contains gateway calls and processing for characteristics. Here's and example:

```
module.exports = [
  { method: 'GET', path:
'/nodes/{node_id}/services/{service_id}/characteristics/{characteristic_id}
', config: { auth: 'simple', handler: handler } },
  { method: 'GET', path:
'/nodes/{node_id}/services/{service_id}/characteristics/{characteristic_id}
/notify', config: { auth: 'simple', handler: notifyhandler } },
  { method: 'POST', path:
'/nodes/{node_id}/services/{service_id}/characteristics/{characteristic_id}
', config: { auth: 'simple', handler: postHandler } }
  ]
```

The `management.js` file makes management calls to the Bluetooth gateway. It includes calls for scanning, pairing, unpairing, and resetting, and so on.

The gateway configuration file can found in `~/BluetoothGateway/navible/config/Gateway.js`. This file defines the gateway URL and port. If you want to run web server and Bluetooth gateway server on two different machines, you need to change the URL of the host from `localhost` to the proper IP address:

```
const Config = {
 host: 'localhost', // can be changed to '192.168.1.20',
 port: '8001' // can be changed to '8080'
 };
```

Another important tool for studying the communication between the gateway explorer and the gateway server is provided by the browser. I recommend using Chrome as it is fast and provides good development tools. Press *F12* in the browser, and it will show you the **Inspect** window. Go to the **Network** tab and refresh the home page. It will display all the network calls in one window:

Name	Status	Type	Initiator	Size	Time	Waterfall 1.00 s
192.168.1.111	200	docum...	Other	1.4 KB	121 ms	
style.css	200	stylesh...	(index)	1.2 KB	127 ms	
jquery.js	200	script	(index)	32.9 KB	126 ms	
bootstrap.min.css	200	stylesh...	(index)	(from ...	7 ms	
bootstrap.min.js	200	script	(index)	(from ...	8 ms	
page.js	200	script	(index)	(from ...	9 ms	
require.js	200	script	(index)	6.7 KB	98 ms	
main.js	200	script	(index)	578 B	190 ms	
bootstrap.min.css	200	stylesh...	(index)	(from ...	3 ms	
app.js	200	script	require.js:34	773 B	88 ms	
homePage.js	200	script	require.js:34	1.4 KB	240 ms	
devicePage.js	200	script	require.js:34	4.2 KB	187 ms	
systemPage.js	200	script	require.js:34	768 B	264 ms	
notificationsPage.js	200	script	require.js:34	1.2 KB	212 ms	
ractive.min.js	200	script	require.js:34	53.6 KB	345 ms	
rv.js	200	script	require.js:34	3.7 KB	229 ms	
jquery.js	200	script	require.js:34	32.9 KB	291 ms	
dataHelper.js	200	script	require.js:34	1.5 KB	230 ms	
deviceDetail.html	200	xhr	rv.js:147	2.9 KB	78 ms	
notificationsPage.html	200	xhr	rv.js:147	912 B	127 ms	
deviceList.html	200	xhr	rv.js:147	1.0 KB	99 ms	
systemPage.html	200	xhr	rv.js:147	674 B	142 ms	
events	200	events...	Other	251 B	221 ms	
devices	200	xhr	jquery.js:5	1.0 KB	299 ms	

If you click on the last call, which is **devices**, you will see the Bluetooth gateway devices' response, which is responsible for bringing back the information about all the devices. It is a big JSON payload coming back from the `gateway` server, so I can't post it all here, but it looks like this:

```
✕  Headers  Preview  Response  Timing
▼ {title: "Devices",…}
  ▼ nodes: [{self: {href: "http://localhost:8001/gap/nodes/fa:76:4b:0f:ac:e6"}, handle: "fa764b0face6",…},…]
    ▼ 0: {self: {href: "http://localhost:8001/gap/nodes/fa:76:4b:0f:ac:e6"}, handle: "fa764b0face6",…}
      ▶ AD: [{ADType: "<type1>", ADValue: " <value1>"}]
        address: "fa:76:4b:0f:ac:e6"
      ▼ advertisement: {localName: "PLAYBULB COMET", txPowerLevel: -2,…}
          localName: "PLAYBULB COMET"
        ▶ manufacturerData: {type: "Buffer", data: [77, 73, 80, 79, 87]}
          serviceData: []
        ▶ serviceUuids: ["ff07"]
          txPowerLevel: -2
      ▼ bdaddrs: [{bdaddr: "fa:76:4b:0f:ac:e6", bdaddrType: "public"}]
        ▶ 0: {bdaddr: "fa:76:4b:0f:ac:e6", bdaddrType: "public"}
          connectionState: false
          handle: "fa764b0face6"
          rssi: -68
        ▶ self: {href: "http://localhost:8001/gap/nodes/fa:76:4b:0f:ac:e6"}
          service: []
      ▶ 1: {self: {href: "http://localhost:8001/gap/nodes/a0:18:28:ed:c4:1f"}, handle: "a01828edc41f",…}
      ▶ 2: {self: {href: "http://localhost:8001/gap/nodes/ca:a7:5c:9a:e8:5b"}, handle: "caa75c9ae85b",…}
      ▶ 3: {self: {href: "http://localhost:8001/gap/nodes/c7:a4:59:97:e5:58"}, handle: "c7a45997e558",…}
      ▶ 4: {self: {href: "http://localhost:8001/gap/nodes/c6:f8:d5:d9:b4:2f"}, handle: "c6f8d5d9b42f",…}
      ▶ 5: {self: {href: "http://localhost:8001/gap/nodes/ef:91:6c:d3:55:12"}, handle: "ef916cd35512",…}
      ▶ 6: {self: {href: "http://localhost:8001/gap/nodes/c9:26:94:04:76:c9"}, handle: "c926940476c9",…}
      ▶ 7: {self: {href: "http://localhost:8001/gap/nodes/14:99:e2:1a:9e:ea"}, handle: "1499e21a9eea",…}
        title: "Devices"
```

As you can see, Bluetooth gateway calls are running on `localhost:8001`. For example, the information about node `fa:76:4b:0f:ac:e6` is obtained by calling the following:

```
http://localhost:8001/gap/nodes/fa:76:4b:0f:ac:e6
```

Following the same methodology, you can study the behavior of the gateway explorer as it makes its calls to the Bluetooth gateway.

You can switch this web application with a native mobile application. A good project for practice can be to implement the same gateway explorer in Android/iOS and make it able to talk to the Bluetooth gateway server remotely.

Summary

In this chapter, we were introduced to the Raspberry Pi and discussed how the Raspberry Pi 3 Model B is a great choice for implementing IoT projects. The second part of the chapter was learning how to install Raspbian, a Debian-based Linux operating system, on the Raspberry Pi 3 using NOOBS. Once it was all set up, we updated the software and installed required utilities (such as Node.js and Bluez) on the Raspberry Pi and got it ready for development. After installing the software, we studied how to deploy the Bluetooth SIG's Bluetooth gateway project and got it running with a web server. The Bluetooth gateway communicates with the web server, and we learned how the communication happens between the gateway and the web server. The chapter also extensively referenced Node.js code, as it was our development platform for both the Bluetooth gateway and the web server.

The next chapter will be the last chapter of the book, where we will discuss the future of Bluetooth technology and how it will help reshape the IoT industry.

8
The Future of Bluetooth Low Energy

In this book, we looked at different aspects of Bluetooth Low Energy. Starting with the very basics of Bluetooth Low Energy and the Internet of Things, we covered some projects that informed us about BLE's practicality. At this point, you should be able to implement your own Bluetooth Low Energy projects in your home or at work. In this chapter, we will put our pens down and try to look into the future. It is always tricky business to predict the future of a fast-changing technology. Having said that, I am confident that technology has a certain pattern and that by following the correct path, we should be able to look into the future.

This theoretical chapter will cover the following topics:

- The Internet of Things and the role of Bluetooth Low Energy
- Using the Internet of Things in smart cities and the automobile industry
- Potential research in Bluetooth Low Energy

The Internet of Things and the role of Bluetooth Low Energy

In this section, we will take a quick look at the history and future of technology, with special emphasis on the Internet of Things and Bluetooth Low Energy.

History and its motion

Technology moves forward and unveils one miracle after another. It started in the eighteenth century with the invention of electricity, when Benjamin Franklin did extensive research and connected the dots between lightning and tiny electric charges. While some people do not consider Franklin the inventor of electricity, it is agreed that his research led to the invention of electricity. Later, Nicola Tesla and Thomas Edison put this technology to commercial use. The reason I explicitly mention the invention of commercial electricity is because this invention is responsible for almost all technological advancement humans have made in the last two centuries. Television, radio, calculators, and eventually personal computers are based on electricity.

This is a view of **Eniac**, or the **Electronic Numerical Integrator And Computer**, in Philadelphia, Pennsylvania:

It is very hard to make a distinction between computing and communication. Computers started with the need for standard calculations and management, but the Internet literally converted computers into communication devices. Later, the same technology converted circuit-switched telephony networks to packet-based cellular networks. Today is the age of advanced computing, where terabytes of data flow in the ether from cellular phones and computers through wireless networks.

History moves under a divine guidance.

We are now standing at a juncture where the existing technology has been pushed to its limits. The last revolutionary development on a commercial scale happened in 2007, when Steve Jobs launched the first iPhone. All later smartphones (Android, iOS, and others) were made on the concepts brought to life by Apple. Sure, now we have powerful batteries, better processors, higher RAM, and better cameras, but they haven't caused a revolution of that scale.

The future of technology and the Internet of Things

Every so often, there comes a new platform, product, or service that changes the way we think about the world. Fortunately, the journey of technology is not yet at its peak, and I doubt it will ever be. The next revolution in technology will not come from the computer or cellular phone industry; it will come from elsewhere.

It will come from the Internet of Things.

Augmented reality-based gadgets are one of my prime suspects, as they provide an amazing fusion of the real world with the virtual world. This new way of interaction will give developers and users a whole new technological domain. Unfortunately, the technology is not yet at its peak. It is expensive and only developed for gaming purposes. Once it starts to be developed for a more general audience, with more applications and less bulky hardware, it is going to be a revolutionary change.

According to an estimate by Cisco Systems (`http://www.cisco.com/c/dam/en_us/about/a c79/docs/innov/IoT_IBSG_0411FINAL.pdf`), there will be 50 to 200 billion connected devices by 2020, which makes it a $1.7-trillion industry. The number is huge because it includes the automobile industry, wherein 90% of all the cars produced by 2020 will be connected to the Internet. A glimpse of this can be seen in Tesla and Lucid Motors, who are working on smart cars. The following shows Lucid Air (a luxury car by Lucid Motors), which will be available by the end of 2018:

"Internet of Things" is a very broad term; just like Internet, it can refer to what the Internet actually is (a network) or something that the Internet does (for example, providing the ability to connect people, send messages, or make a voice call). The Internet of Things can be used to refer to:

- A smart car (for example, Lucid Air or Tesla Model 3)
- Smart devices in your home (for example, Google Home or Phillips Hue Lighting)

- Controlling your home devices remotely through your cellular phone (for example, controlling them using a Bluetooth gateway)
- Remote monitoring

Bluetooth Low Energy in the future of IoT

Bluetooth has always been a very attractive technology in consumer electronics. When Bluetooth was integrated with cellular phones, the adoption rate of this technology increased tremendously. It completely replaced infrared communications as that required line of sight to transfer data. On the other hand, Bluetooth was a cheap and easily integrated solution that did not require line of sight and could transfer data at higher speed. The result was obvious: every cellphone and laptop manufacturer started to integrated Bluetooth in their devices, and now, Bluetooth is probably the most widely available short-distance communication technology in the market. As of now, almost all cell phones and laptops come with built-in Bluetooth for data transfer and multimedia use(wireless earbuds/headphones).

A detailed history of Bluetooth was discussed in the first chapter, but in this section, we will emphasize the creation of Bluetooth Low Energy. Although Bluetooth was a widely adopted technology, it was still power hungry. The Bluetooth SIG realized this and reinvented the technology with a special focus on power efficiency. Bluetooth version 4.0 was made with a completely new protocol stack. You can find more information about this in `Chapter 1`, *BLE and the Internet of Things*.

Now that Bluetooth Low Energy is power efficient with increased data rates and range, it is the perfect technology to become a part of the Internet of Things. It already has a strong market and brand recognition throughout the world, and it is believed that BLE and WLAN are going to be an important part of the future. In 2016, all major top-of-the-line phones came with Bluetooth Low Energy v4.2, and it will be carried forward with integration into Bluetooth 5 in 2017. The effect of this high adoption rate will surely be felt on the Internet of Things. Since BLE is a cheap and power-efficient technology, it can easily be integrated with appliances and smart-home devices. Refrigerators, lights, motion sensors, coffee machines, water bottles, and many other things can be made smart just by using a BLE chip. Once Bluetooth Low Energy is widely adopted by new companies to produce smart *things*, they can accept and send commands to cellular phones.

The competitor of Bluetooth Low Energy is Wireless Local Area Network, which has the same adoption rate as Bluetooth. It also has faster speed, greater range, and the ability to talk directly to the Internet, but Bluetooth is catching up. With the launch of Bluetooth 5, the SIG announced that they are extending the range of Bluetooth by four times and speed twice. They also announced the inclusion of a Bluetooth mesh with more secure communication. The Bluetooth SIG made sure that the new version of Bluetooth will work with less interference and improved interoperability with other wireless technologies. These are the exact things Bluetooth needs in order to compete with other technologies on the platform of the Internet of Things. A visualization of BLE working in a big IoT cloud follows:

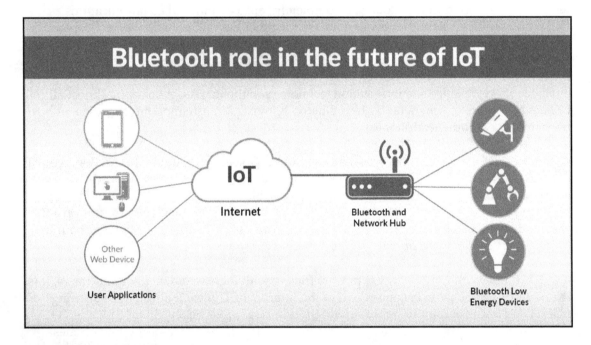

With this, Bluetooth beacons are even cheaper and low cost; their only job is to broadcast data instead of receiving general-purpose Bluetooth commands. We can take Estimote Nearables as an example, which are like stickers but can provide Bluetooth broadcasts for over a year. On the other hand, Estimote Location Beacons, which are sturdy and come with many sensors, can make Bluetooth broadcasts for over 5 years. This shows that the technology is power efficient and can help in the expansion of Bluetooth in the IoT.

According to ABI, 48 billion Internet-enabled devices will support Bluetooth by 2021, which makes it around a third of the total Internet of Things market. As mentioned before, Bluetooth will coexist with fellow technologies such as Wi-Fi and LTE, and it will provide a more robust connection and stability to the whole IoT ecosystem. The new version of Bluetooth also has a simpler architecture, reduced cost, and greater range and speed, all of which increase its use cases and provide new opportunities for companies to use it in innovative ways.

Use of IoT in smart cities and the automobile industry

Smart cities are cities created with the vision of combining information and communication technologies. This is achieved by including IoT solutions to manage the city. Management implies the management of a city's assets, for example power plants, healthcare systems, transportation systems, schools, libraries, sewage systems, waste management systems, and law enforcement systems. The Internet of Things is used to enhance the quality and ease of interactivity. The goal is to improve the quality of the city by automating various aspects of life in a city. Smart cities don't just emphasize automation, but also try to increase the quality of communication by providing new means.

 According to research, the smart city market will be an industry worth hundreds of billions of dollars.

By bringing technology into urban life, management can be made easy. It can also improve the relationship between the people and the city's authorities. A smart city can also be referred to as an information city, a knowledge-based city, telecity, teletopia, mesh city, cyberville, digital city, electronic communities, wired city, and so on. Many major cities are in the fight to achieve the status of a smart city. Some examples of smart cities are Milton Keynes, Southampton, Laguna Croata, Manchester, New Songdo City, Santa Cruz, Singapore, Amsterdam, Barcelona, Madrid, and Stockholm. On the other hand, Tel Aviv, Israel was awarded the World Smart City Award in 2014.

 There are six aspects of smart cities: living, environment, people, governance, mobility, and economy. Smart cities needs to satisfy each of these aspects. This means that for smart cities, all decisions (either technical or financial) should be taken by considering all these six aspects.

An abstract diagram depicting the vision of smart cities is shown here:

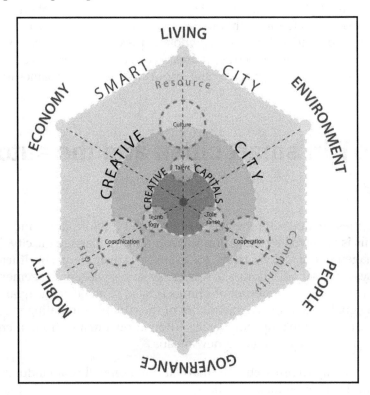

Technical challenges

One of the potential issues in the adoption of smart cities is the non-interoperability of the divergent technologies. As of now, you can find many technologies working independently in urban environments. This will be a challenge to handle, and only IoT's vision can become the base here. It has the capability of unifying information and communication platforms, which makes it a strong candidate for smart cities. As of 2017, a strong technical model for a unified Internet of Things is still lacking, but as the technology is evolving, we can expect it to converge into one big IoT bubble.

Another important technical challenge is to make IoT simple for the masses. The adoption rate for IoT technology is not as fast as it can be. The vision of it being a trillion-dollar industry will not be realizable until it is easy to use. It will be a challenge to convince developing countries to adopt the concept of smart cities.

Financial challenges

From the financial standpoint, no clear model is being developed for smart cities. The economic conditions of the public sector are worsening, and not all countries have money to spend on smart cities. Developing countries are at a higher risk than developed countries. This limitation prevents an exponential adoption of the smart-city market. A possible solution of this problem is given in a research paper by Andrea Zanella et al:

> *"A possible way out of this impasse is to first develop those services that conjugate social utility with very clear return on investment, such as smart parking and smart buildings, and will hence act as catalyzers for the other added-value services."*

Potential research in Bluetooth Low Energy

Research and development in Bluetooth Low Energy is a continuing process undertaken by the Bluetooth SIG. Many companies, in collaboration with the SIG, are working to stretch the limits of the technology.

To understand the scale of research and development, you can refer to the list of contributors to Bluetooth 5:

Amre El-Hoiydi Phonak Communications AG	Angel Polo Broadcom	Anthony Viscardi Texas Instruments
Bjarne Klemmensen Oticon A/S	Brian A. Redding Qualcomm Technologies, Inc.	Burch Seymour Continental Automotive
Chris Deck ON Semiconductor	Clive D.W. Feather Samsung Electronics Co., Ltd.	David Engelien-Lopes Nordic Semiconductor ASA
Dishant Srivastava CSR	Edward Harrison Anritsu	Eivind Sjøgren Olsen Nordic Semiconductor ASA
Florian Lefeuvre Texas Instruments	Harish Balasubramaniam Intel	Huanchun Ye Broadcom
James Wang MediaTek	Jean-Philippe Lambert RivieraWaves	Jeff Solum Starkey Hearing Technologies
Joel Linsky Qualcomm Technologies, Inc.	Johan Hedberg Intel	Jonathan Tanner Qualcomm Technologies, Inc.
Josselin de la Broise Marvell	KC Chou MediaTek	Knut Odman Broadcom
L.C. Ko MediaTek	Laurence Richardson Qualcomm Technologies, Inc.	Marcel Holtmann Intel
Mayank Batra Qualcomm Technologies, Inc.	Michael Knudsen Samsung Electronics Co., Ltd.	Michael Ungstrup Widex A/S
Niclas Granqvist Polar	Phil Corbishley Nordic Semiconductor ASA	Phil Hough Anritsu
Raja Banerjea CSR	Rasmus Abildgren Samsung Electronics Co., Ltd.	RaviKiran Gopalan Qualcomm Atheros
Robert Hulvey Broadcom	Robin Heydon Qualcomm Technologies, Inc.	Sam Geeraerts NXP
Sandipan Kundu CSR	Shawn Ding Broadcom	Steven Hall Broadcom
Thomas Varghese Mindtree	Till Schmalmack Phonak Communications AG	Tim Wei IVT Wireless
Tomás Motos López Texas Instruments	Yi-Ling Chao Marvell	

This list is available on the Bluetooth SIG's website. You can see that a broad range of people contributed to the development of Bluetooth 5. The list goes from Qualcomm Technologies, Inc. to Texas Instruments. With their collaboration, they managed to achieve extraordinary advancements in Bluetooth technology. The proof of this is the significant increase in the range and speed of the latest version of Bluetooth.

The SIG also encourages research from students and independent research companies to improve the health of Bluetooth technology. In this section, we will discuss some potential research topics that students/companies can put their efforts into.

Enhancing meshes in Bluetooth

Meshes are a relatively new domain in Bluetooth, and just like wireless mesh networks, they will go through some serious revisions. Since a mesh is a fast-changing wireless network, there is a need for a universal routing algorithm that is smart enough to configure itself intelligently and avoid any denial-of-service attacks. It will also be a challenge for Bluetooth to acknowledge other IoT standards and provide a sophisticated mesh that is able to connect to other mesh networks. This is only possible if researchers take into consideration the wide variety of standards in mesh networking. **Adhoc On-demand Distance Vector** (**AODV**) and **Dynamic Source Routing** (**DSR**) are two basic algorithms for routing in ad-hoc networks. Bluetooth requires a gateway protocol that can easily route to these protocols.

We don't know the official Bluetooth mesh-routing algorithm yet, but there is a possibility that the routing will be done via **flooding**. Flooding is a routing algorithm where every incoming packet is sent through every outgoing link. If adopted, this approach for routing can involve serious security threats. Independent research can be conducted to come up with efficient routing algorithms for Bluetooth mesh technology.

Enhancing security in Bluetooth

Bluetooth Low Energy v4.0 and 4.1 used LE legacy pairing, which is similar to BR/EDR secure simple pairing and does not provide passive eavesdropping protection. The issue was later resolved in Bluetooth v4.2 and 5, which use the *Diffie-Hellman Elliptic Curve* algorithm for key exchange prior to pairing. The whole link is then encrypted for any further communication, avoiding any passive eavesdropping.

Bluetooth does provide a good security mechanism for paired devices, but it is still unclear how to restrict sniffers from investigating the services and characteristics of any Bluetooth Low Energy device. If you download the Nordic Connect Android/iOS application, you will be able to scan Bluetooth devices and read their characteristics. Much research can be performed to secure Bluetooth by not letting anyone sniff and investigate services/characteristics of Bluetooth devices.

Bluetooth Low Energy does not consider many-to-many relationships. The new versions are good at encrypting one-to-one links, where one central is paired with one peripheral, but the technology does not provide any mechanism to successfully pair multiple centrals with multiple peripherals. As of now, developers need to implement their own security wrapper to accommodate this scenario.

> Refer to `Chapter 1`, *BLE and the Internet of Things*, for complete details on security mechanisms for Bluetooth Low Energy.

Mesh networks are vulnerable to many security threats, and Bluetooth mesh is no different. Mesh network attacks can be broadly classified into two categories:

- Insider attacks
- Outsider attacks

Complete details of these attacks are given in `Chapter 6`, *Bluetooth Mesh Technology*. In a nutshell, insider attacks are when an intruder is a part of a mesh network, and outsider attacks are when an intruder has remote access to the network but is not physically present inside. Traditionally, firewalls can be used as a prime source of security from outsider attacks in wireless mesh networks, and this is only possible when you are using this network in a private environment. Since Bluetooth promises its role in IoT, mesh networking needs to come with proper security.

For most meshes, insider attacks are even more threatening, as any intruder can act as a normal node of a mesh. Intrusion detection and prevention systems are required to fight against insider attacks, and only then will it be possible to put Bluetooth mesh technology on the map of the industrial world. These topics open many doors for future research in the field of Bluetooth.

Summary

This is the closing chapter of this book, in which I have given you a glimpse into the future. The discussions about the future of the technology are ongoing, so it was almost impossible to cater to every possible future scenario. We briefly went through the history of technology and the future of IoT, with a special focus on the role of Bluetooth Low Energy. We also discussed technical and financial challenges, risks of Bluetooth mesh networks, and potential research scenarios for Bluetooth Low Energy security.

This is an open invitation for everyone to think how you can get the most out of this technology. Implement home projects, and take the initiative in converting your home into a smart home. Do not wait for smart cities to come to you; rather, make your home smart. There is much room for improvement in the technology, which can only be achieved by putting effort on small and big scales alike. If you think that Bluetooth does not provide some feature out of the box, develop it and share the resource with the community. In the end, technology is best used for the benefit of humankind.

> *"Technology is nothing. What's important is that you have a faith in people, that they're basically good and smart, and if you give them tools, they'll do wonderful things with them."*
>
> *-Steve Jobs*

Index